PENGUIN BOOKS

Optimism over Despair

Noam Chomsky was born in Philadelphia, Pennsylvania, in 1928. He is the bestselling author of over a hundred influential political books, including *Who Rules the World?*, *Hegemony or Survival*, *Imperial Ambitions*, *Failed States*, *Interventions*, *What We Say Goes*, *Hopes and Prospects*, *Making the Future*, *Gaza in Crisis*, *Occupy*, *Power Systems*, *On Anarchism*, *Because We Say So* and *Masters of Mankind*. He has also been the subject of numerous books of biography and interview, and has collaborated with journalists on books including *Perilous Power*, *Gaza in Crisis* and *On Palestine*. Noam Chomsky is Professor Emeritus at Massachusetts Institute of Technology (MIT).

Optimism over Despair

On Capitalism, Empire and Social Change

Noam Chomsky
and C. J. Polychroniou

PENGUIN BOOKS

PENGUIN BOOKS

UK | USA | Canada | Ireland | Australia
India | New Zealand | South Africa

Penguin Books is part of the Penguin Random House group of companies
whose addresses can be found at global.penguinrandomhouse.com.

First published in the United States of America by Haymarket Books 2017
First published in Great Britain by Penguin Books 2017
001

This book was published with the generous support of
Lannan Foundation and Wallace Action Fund

Printed in Great Britain by Clays Ltd, St Ives plc

A CIP catalogue record for this book is available from the British Library

ISBN: 978-0-241-98197-9

www.greenpenguin.co.uk

Contents

Introduction

The interviews in this volume present the views of the world's leading public intellectual on the consequences of capitalist globalization, and much more, as recorded in conversations with the undersigned over the course of the last four years—from late 2013 to early 2017, to be exact—and originally published in *Truthout*.

Noam Chomsky has been "America's moral conscience" for more than half a century (even if he remains unknown to the majority of Americans) as well as the world's most recognized public intellectual, consistently speaking out against US aggression and defending the rights of the weak and the oppressed throughout the world from the time of the Vietnam War to the present. His analyses are always grounded in indisputable facts and are also guided by deeply held moral considerations about freedom, democracy, human rights, and human decency.

Chomsky's voice remains almost singularly a beacon of hope and optimism in these dark times—an age of unparalleled economic inequality, growing authoritarianism, and social Darwinism, with a left that has turned its back on the class struggle.

For quite some time now, there have been clear and strong indications across the entire political and socioeconomic spectrum in advanced Western societies that the contradictions of capitalist globalization and the neoliberal policies associated with them threaten to unleash powerful forces with the capacity to produce not only highly destructive outcomes for growth and prosperity, justice, and social peace, but also concomitant consequences for democracy, the environment, and human civilization on the whole.

Still, according to Chomsky, despair is not an option. No matter how horrendous the current world situation appears to be, resistance to oppression and exploitation has never been a fruitless undertaking, even in darker times than our own. Indeed, the Trump "counterrevolution" in the United States has already brought to the surface a plethora of social forces determined to stand up to the aspiring autocrat, and the future of resistance in the world's most powerful country appears more promising than in many other parts of the advanced industrialized world.

In this context, the interviews assembled here are, we believe, of critical import. These were originally commissioned and edited by Maya Schenwar, Alana Yu-lan Price, and Leslie Thatcher for publication as stand-alone articles in *Truthout*. Our hope in anthologizing them is that they will assist to introduce the views and ideas of Noam Chomsky to a new generation of readers, while maintaining faith among the rest in the human ability to provide tenacious resistance to the forces of political darkness and ultimately change the course of history for the better.

—C. J. Polychroniou, March 2017

Part I

The Breakdown of American Society and a World in Transition

C. J. POLYCHRONIOU: Noam, you have said that the rise of Donald Trump is largely due to the breakdown of American society. What exactly do you mean by this?

NOAM CHOMSKY: The state-corporate programs of the past thirty-five or so years have had devastating effects on the majority of the population, with stagnation, decline, and sharply enhanced inequality being the most direct outcomes. This has created fear and has left people feeling isolated, helpless—victims of powerful forces they can neither understand nor influence. The breakdown is not caused by economic laws. They are policies, a kind of class war initiated by the rich and powerful against the working population and the poor. This is what defines the neoliberalism period, not only in the United States but in Europe and elsewhere. Trump is appealing to those who sense and experience the breakdown of American society—to deep feelings of anger, fear, frustration, hopelessness, probably among sectors of the population that are seeing an increase in mortality, something unheard of apart from war.

Class warfare remains as vicious and one-sided as ever. Neoliberal governance over the last thirty years, regardless if there was a Republican or a Democratic administration in place, has intensified immensely the processes of exploitation and induced ever-larger gaps between haves and have-nots in American society. Moreover, I don't see neoliberal

class politics being on retreat in spite of the opportunities that opened up because of the last financial crisis and by having a centrist Democrat in the White House.

The business classes, which largely run the country, are highly class conscious. It is not a distortion to describe them as vulgar Marxists, with values and commitments reversed. It was not until thirty years ago that the head of the most powerful union recognized and criticized the "one-sided class war" that is relentlessly waged by the business world. It has succeeded in achieving the results you describe. However, neoliberal policies are in shambles. They have come to harm the most powerful and privileged (who only partially accepted them for themselves in the first place), so they cannot be sustained.

It is rather striking to observe that the policies that the rich and powerful adopt for themselves are the precise opposite of those they dictate to the weak and poor. Thus, when Indonesia has a deep financial crisis, the instructions from the US Treasury Department (via the International Monetary Fund, IMF) are to pay off the debt (to the West), to raise interest rates and thus slow the economy, to privatize (so that Western corporations can buy up their assets), and the rest of the neoliberal dogma. For ourselves, the policies are to forget about debt, to reduce interest rates to zero, to nationalize (but not to use the word), and to pour public funds into the pockets of the financial institutions, and so on. It is also striking that the dramatic contrast passes unnoticed, along with the fact that this conforms to the record of the economic history of the past several centuries, a primary reason for the separation of the first and third worlds.

Class politics is so far only marginally under attack. The Obama administration has avoided even minimal steps to end and reverse the attack on unions. Obama has even indirectly indicated his support for this attack, in interesting ways. It is worth recalling that his first trip to show his solidarity with working people (called "the middle class," in US rhetoric) was to the Caterpillar plant in Illinois. He went there in defiance of pleas by church and human rights organizations because of Caterpillar's grotesque role in the Israeli occupied territories, where it is a prime instrument in devastating the land and villages of "the wrong people." But it seems not even to have been noticed that, adopting Reagan's antilabor policies, Caterpillar became the first industrial corporation in generations to break a powerful union by employing strike-breakers, in radical violation of international labor conventions. That left the United States alone in the industrial world, along with apartheid South Africa, in tolerating such means of undermining workers' rights and democracy—and now I presume the United States is alone. It is hard to believe that the choice was accidental.

There is a widespread belief, at least among some well-known political strategists, that issues do not define American elections—even if the rhetoric is that candidates need to understand public opinion in order to woo voters—and we do know, of course, that media provide a wealth of false information on critical issues (take the mass media's role before and during the launching of the Iraq War) or fail to provide any information at all (on labor issues, for example). Yet, there is strong evidence indicating that the American public cares about the great social, economic, and foreign policy issues facing the country. For example, according to a research study released some years ago by the University of Minnesota, Americans ranked health care among the most important problems facing the country. We also know that the overwhelming majority of Americans are in support of unions. Or that they judged the "war against terror" to be a total failure. In the light of all of this, what's the best way to understand the relation between media, politics, and the public in contemporary American society?

It is well established that electoral campaigns are designed so as to marginalize issues and focus on personalities, rhetorical style, body language, and the like. And there are good reasons. Party managers read polls and are well aware that on a host of major issues, both parties are well to the right of the population—not surprisingly; they are, after all, business parties. Polls show that a large majority of voters object, but those are the only choices offered to them in the business-managed electoral system, in which the most heavily funded candidate almost always wins.

Similarly, consumers might prefer decent mass transportation to a choice between two automobiles, but that option is not provided by advertisers—nor, indeed, by markets. Ads on TV do not provide information about products; rather, they provide illusion and imagery. The same public relations firms that seek to undermine markets by ensuring that uninformed consumers will make irrational choices (contrary to abstract economic theories) seek to undermine democracy in the same way. And the managers of the industry are well aware of all of this. Leading figures in the industry have exulted in the business press that they have been marketing candidates like commodities ever since Reagan, and this is their greatest success yet, which they predict will provide a model for corporate executives and the marketing industry in the future.

You mentioned the Minnesota poll on health care. It is typical. For decades, polls have shown that health care is at or near the top of public concerns—not surprisingly, given the disastrous failure of the health care system, with per capita costs twice as high as comparable societies and some of the worst outcomes. Polls also consistently show that large majorities want a nationalized system, called "single payer," rather like the existing Medicare system for the elderly, which is far more efficient than the privatized systems or the one introduced by Obama. When any of this is mentioned, which is rare, it is called "politically impossible" or "lacking political support"—meaning that the insurance and pharmaceutical industries, and others who benefit from the current system, object. We gained an interesting insight into the workings of American democracy from the fact that in 2008, unlike in 2004, the Democratic candidates—first Edwards, then Clinton and Obama—came forward with proposals that at least began to approach what the public has wanted for decades. Why? Not because of a shift in public attitudes, which have remained steady. Rather, the manufacturing industry has been suffering from the costly and inefficient privatized health care system, and the enormous privileges granted, by law, to the pharmaceutical industries. When a large sector of concentrated capital favors some program, it becomes "politically possible" and has "political support." Just as revealing as the facts themselves is that they are not noticed.

Much the same is true on many other issues, domestic and international.

The US economy is facing myriad problems, although profits for the rich and corporations returned long ago to the levels they were prior to the eruption of the 2008 financial crisis. But the one single problem that most academic and financial analysts seem to focus on as being of most

critical nature is that of government debt. According to mainstream analysts, US debt is already out of control, which is why they have been arguing consistently against big economic stimulus packages to boost growth, contending that such measures will only push the United States deeper into debt. What is the likely impact that a ballooning debt will have on the American economy and on international investors' confidence in the event of a new financial crisis?

No one really knows. Debt has been far higher in the past, particularly after World War II. But that was overcome thanks to the remarkable economic growth under the wartime semi-command economy. So we know that if government stimulus spurs sustained economic growth, the debt can be controlled. And there are other devices, such as inflation. But the rest is very much guesswork. The main funders—primarily China, Japan, oil producers—might decide to shift their funds elsewhere for higher profits. But there are few signs of such developments, and they are not too likely. The funders have a considerable stake in sustaining the US economy for their own exports. There is no way to make confident predictions, but it seems clear that the entire world is in a tenuous situation, to say the least.

You seem to believe, in contrast to so many others, that the United States remains a global economic, political, and of course military superpower even after the latest crisis—and I do have the same impression, as well, as the rest of the world economies are not only not in any shape to challenge America's hegemony but are looking toward the United States as a savior of the global economy. What do you see as the competitive advantages that US capitalism has over the EU economy and the newly emerging economies in Asia?

The 2007–2008 financial crisis in large measure originated in the United States, but its major competitors—Europe and Japan—ended up suffering more severely, and the United States remained the choice location for investors who are looking for security in a time of crisis. The advantages of the United States are substantial. It has extensive internal resources. It is unified, an important fact. Until the Civil War in the 1860s, the phrase "United States" was plural (as it still is in European languages). But since then, the phrase has been singular, in standard English. Policies designed in Washington by state power and concentrated capital apply to the whole country. That is far harder in Europe. A couple of years after the eruption of the latest global financial crisis, the European Commission task force issued a report saying, "Europe needs new bodies to monitor systemic risk and coordinate oversight of financial institutions across the region's patchwork of supervision," though the task force, headed then by a former French central banker, "stopped well short of suggesting a single European watchdog"—which the United States can have any time it wants. For Europe, it would be "an almost impossible mission," the task force leader said. [Several] analysts, including the *Financial Times*, have described such a goal as politically impossible, "a step too far for many member states reluctant to cede authority in this area." There are many other advantages to unity. Some of the harmful effects of European inability to coordinate reactions to the crisis have been widely discussed by European economists.

The historical roots of these differences between Europe and the United States are familiar. Centuries of conflict imposed a nation-state system in Europe, and the experience of World War II convinced Europeans that they must abandon their traditional sport of slaughtering one another, because the next try would be the last. So we have what political scientists like to call "a democratic peace," though it is far from clear that democracy has much to do with it. In contrast, the United States is a settler-colonial state, which murdered the indigenous population and consigned the remnants to "reservations," while conquering half of Mexico, then expanding beyond. Far more than in Europe, the rich internal diversity was destroyed. The Civil War cemented central authority, and uniformity in other domains as well: national language, cultural patterns, huge state-corporate social engineering projects such as the suburbanization of the society, massive central subsidy of advanced industry by research and development, procurement and other devices, and much else.

The new emerging economies in Asia have incredible internal problems unknown in the West. We know more about India than China, because it is a more open society. There are reasons why it ranks 130th in the Human Development Index (about where it was before the partial neoliberal reforms); China ranks 90th, and the rank could be worse if more were known about it. That only scratches the surface. In the eighteenth century, China and India were the commercial and industrial centers of the world, with sophisticated market systems, advanced health levels by comparative standards, and so on. But imperial conquest and economic policies (state intervention for the rich, free markets rammed down the throats of the poor) left them in miserable conditions. It is notable that the one country of the global South that developed was Japan, the one country that was not colonized. The correlation is not accidental.

Is the United States still dictating IMF policies?

It's opaque, but my understanding is that IMF's economists are supposed to be, maybe are, somewhat independent of the political people. In the case of Greece, and austerity generally, the economists have come out with some strongly critical papers on the Brussels programs, but the political people seem to be ignoring them.

On the foreign policy front, the "war on terror" seems to be a never-ending enterprise and, as with the Hydra monster, two new heads pop up when one is cut off. Can massive interventions of force wipe out terrorist organizations like ISIS (also known as Daesh or ISIL)?

Upon taking office, Obama expanded intervention forces and stepped up the wars in Afghanistan and Pakistan, just as he had promised he would do. There were peaceful options, some recommended right in the mainstream: in *Foreign Affairs*, for example. But these did not fall under consideration. Afghan president Hamid Karzai's first message to Obama, which went unanswered, was a request to stop bombing civilians. Karzai also informed a UN delegation that he wanted a timetable for withdrawal of foreign (meaning US) troops. Immediately he fell out of favor in Washington, and accordingly shifted from a media favorite to "unreliable," "corrupt," and so on—which was no more true than when he was feted as our "our man" in Kabul. Obama sent many more troops and stepped up bombing on both sides of the Afghan–Pakistan border—the Durand line, an artificial border established by the British, which cuts the Pashtun areas in two and which the people have never accepted. Afghanistan in the past often pressed for obliterating it.

That is the central component of the "war on terror." It was certain to stimulate terror, just as the invasion of Iraq did, and as resort to force does quite generally. Force can succeed. The existence of the United States is one illustration. The Russians in Chechnya is another. But it has to be overwhelming, and there are probably too many tentacles to wipe out the terrorist monster that was largely created by Reagan and his associates, since nurtured by others. ISIS is the latest one, and a far more brutal organization than al-Qaeda. It is also different in the sense that it has territorial claims. It can be wiped out through massive employment of troops on the ground, but that won't end the emergence of similar-minded organizations. Violence begets violence.

US relations with China have gone through different phases over the past few decades, and it is hard to get a handle on where things stand today. Do you anticipate future US–Sino relations to improve or deteriorate?

The US has a love-hate relation with China. China's abysmal wages, working conditions, and lack of environmental constraints are a great boon to US and other Western manufacturers who transfer operations there, and to the huge retail industry, which can obtain cheap goods. And the United States now relies on China, Japan, and others to sustain its own economy. But China poses problems as well. It does not intimidate easily. When the United States shakes its fist at Europe and tells Europeans to stop doing business with Iran, they mostly comply. China doesn't pay much attention. That's frightening. There is a long history of conjuring up imaginary Chinese threats. It continues.

Do you see China being in a position any time soon to pose a threat to US global interests?

Among the great powers, China has been the most reserved in use of force, even military preparations. So much so that leading US strategic analysts (John Steinbrunner and Nancy Gallagher, writing in the journal of the ultra-respectable *American Academy of Arts and Sciences*) called on China some years ago to lead a coalition of peace-loving nations to confront the US aggressive militarism that they think is leading to "ultimate doom."

There is little indication of any significant change in that respect. But China does not follow orders and is taking steps to gain access to energy and other resources around the world. That constitutes a threat.

Indian–Pakistani relations pose clearly a major challenge in US foreign policy. Is this a situation the United States can actually have under control?

To a limited extent. And the situation is highly volatile. There is constant ongoing violence in Kashmir—state terror by India, Pakistan-based terrorists. And much more, as the recent Mumbai bombings revealed. There are also possible ways to reduce tensions. One is a planned pipeline to India through Pakistan from Iran, the natural source of energy for India. Presumably, Washington's decision to undermine the nonproliferation treaty by granting India access to nuclear technology was in part motivated by the hope of undercutting this option and bringing India to join in Washington's campaign against Iran. It also may be a related issue in Afghanistan, where there has long been discussion of a pipeline (TAPI) from Turkmenistan through Afghanistan to Pakistan and then India. It is probably not a very live issue, but quite possibly is in the background. The "great game" of the nineteenth century is alive and well.

In many circles, there is a widespread impression that the Israel lobby calls the shots in US foreign policy in the Middle East. Is the power of the Israel lobby so strong that it can have sway over a superpower?

My friend Gilbert Achcar, a noted specialist on the Middle East and international affairs generally, describes that idea as "phantasmagoric." Rightly. It is not the lobby that intimidates US high-tech industry to expand its investments in Israel, or that twists the arm of the US government so that it will pre-position supplies there for later US military operations and intensify close military and intelligence relations.

When the lobby's goals conform to perceived US strategic and economic interests, it generally gets its way: crushing of Palestinians, for example, a matter of little concern to US state-corporate power. When goals diverge, as often happens, the lobby quickly disappears, knowing better than to confront authentic power.

I agree totally with your analysis, but I think you would also agree that the Israel lobby is influential enough, and beyond whatever economic and political leverage it carries, that criticisms of Israel still cause hysterical reactions in the United States—and you certainly have been a target of right-wing Zionists for many years. To what do we attribute this intangible influence on the part of the Israel lobby over American public opinion?

That is all true, though much less so than in recent years. It is not really power over public opinion. In numbers, by far the largest support for Israeli actions is independent of the lobby: Christian religious fundamentalists. British and American Zionism preceded the Zionist movement, based on providentialist interpretations of Biblical prophecies. The population at large supports the two-state settlement, doubtless unaware that the United States has been unilaterally blocking it. Among educated sectors, including Jewish intellectuals, there was little interest in Israel before its great military victory in 1967, which really established the US–Israeli alliance. That led to a major love affair with Israel on the part of the educated classes. Israel's military prowess and the US–Israeli alliance provided an irresistible temptation to combine support for Washington with worship of power and humanitarian pretexts. But to put it in perspective, reactions to criticism of US crimes are at least as severe, often more so. If I count up the death threats I have received over the years, or the diatribes in journals of opinion, Israel is far from the leading factor. The phenomenon is by no means restricted to the United States. Despite much self-delusion, Western Europe is not very different—though, of course, it is more open to criticism of US actions. The crimes of others usually tend to be welcome, offering opportunities to posture about one's profound moral commitments.

Under Erdoğan, Turkey has been in a process of unfolding a neo-Ottoman strategy towards the Middle East and Central Asia. Is the unfolding of this grand strategy taking place with the collaboration or the opposition of the United States?

Turkey, of course, has been a very significant US ally, so much so that under Clinton it became the leading recipient of US arms (after Israel and Egypt, in a separate category). Clinton poured arms into Turkey to help it carry out a vast campaign of murder, destruction, and terror against its Kurdish minority.

Turkey has also been a major ally of Israel since 1958, part of a general alliance of non-Arab states, under the US aegis, with the task of ensuring control over the world's major energy sources by protecting the ruling dictators against what is called "radical nationalism"—a euphemism for the populations. US–Turkish relations have sometimes been strained. That was particularly true in the buildup to the US invasion of Iraq, when the Turkish government, bowing to the will of 95 percent of the population, refused to join. That caused fury in the United States. Paul Wolfowitz was dispatched to order the disobedient government to mend its evil ways, to apologize to the United States, and to recognize that its duty is to help the United States. These well-publicized events in no way undermined Wolfowitz's reputation in the liberal media as the "idealist in chief" of the Bush administration, utterly dedicated to promoting democracy. Relations are somewhat tense today too, though the alliance is in place. Turkey has quite natural potential relations with Iran and Central Asia and might be inclined to pursue them, perhaps raising tensions with Washington again. But it does not look too likely right now.

On the Western front, are plans for the eastward expansion of NATO, which go back to the era of Bill Clinton, still in place?

One of Clinton's major crimes in my opinion—and there were many—was to expand NATO to the east, in violation of a firm pledge to Gorbachev by his predecessors after Gorbachev made the astonishing concession to allow a united Germany to join a hostile military alliance. These very serious provocations were carried forward by Bush, along with a posture of aggressive militarism which, as predicted, elicited strong reactions from Russia. But American redlines are already placed on Russia's borders.

What are your views about the EU? It is still largely a trailblazer for neoliberalism and hardly a bulwark for US aggression. But do you see any signs that it can emerge at some point as a constructive, influential actor on the world stage?

It could. That is a decision for Europeans to make. Some have favored taking an independent stance, notably De Gaulle. But by and large, European elites have preferred passivity, following pretty much in Washington's footsteps.

Horror Beyond Description: The Latest Phase of the "War on Terror"

C. J. POLYCHRONIOU: I would like to start by hearing your thoughts on the latest developments on the war against terrorism, a policy that dates back to the Reagan years and was subsequently turned into a doctrine of Islamophobic "crusade" by George W. Bush, with simply inestimable cost to innocent human lives and astonishingly profound effects for international law and world peace. The war against terrorism is seemingly entering a new and perhaps more dangerous phase as other countries have jumped into the fray, with different policy agendas and interests than those of the United States and some of its allies. First, do you agree with the above assessment on the evolution of the war against terrorism, and, if so, what are likely to be the economic, social, and political consequences of a permanent global war on terror, for Western societies in particular?

NOAM CHOMSKY: The two phases of the "war on terror" are quite different, except in one crucial respect. Reagan's war very quickly turned into murderous terrorist wars, presumably the reason why it has been "disappeared." His terrorist wars had hideous consequences for Central America, southern Africa, and the Middle East. Central America, the most direct target, has yet to recover, one of the primary reasons—rarely mentioned—for

---

Originally published in *Truthout*, December 3, 2015

the current refugee crisis. The same is true of the second phase, redeclared by George W. Bush twenty years later, in 2001. Direct aggression has devastated large regions, and terror has taken new forms, notably Obama's global assassination (drone) campaign, which breaks new records in the annals of terrorism, and, like other such exercises, probably generates dedicated terrorists more quickly than it kills suspects.

The target of Bush's war was al-Qaeda. One hammer blow after another—Afghanistan, Iraq, Libya, and beyond—has succeeded in spreading jihadi terror from a small tribal area in Afghanistan to virtually the whole world, from West Africa through the Levant and on to Southeast Asia. One of history's great policy triumphs. Meanwhile, al-Qaeda has been displaced by much more vicious and destructive elements. Currently, ISIS holds the record for monstrous brutality, but other claimants for the title are not far behind. The dynamic, which goes back many years, has been studied in an important work by military analyst Andrew Cockburn, in his book *Kill Chain*. He documents how when you kill one leader without dealing with the roots and causes of the phenomenon, he is typically replaced very quickly by someone younger, more competent, and more vicious.

One consequence of these achievements is that world opinion regards the United States as the greatest threat to peace by a large margin. Far behind, in second place, is Pakistan, presumably inflated by the Indian vote. Further successes of the kind already registered might even create a broader war with an inflamed Muslim world while the Western societies subject themselves to internal repression and curtailing of civil rights and groan under the burden of huge expenses, realizing Osama bin Laden's wildest dreams, and those of ISIS today.

In US policy discussions revolving around the "war on terror," the difference between overt and covert operations has all but disappeared. Meanwhile the identification of terrorist groups and the selection of actors or states supporting terrorism not only appear to be totally arbitrary, but also in some cases the culprits identified have raised questions about whether the "war on terror" is in fact a real war against terrorism or whether it is a smokescreen to justify policies of global conquest. For example, while al-Qaeda and ISIS are undeniable terrorist and murderous

organizations, the fact that US allies such as Saudi Arabia and Qatar, and even NATO member countries such as Turkey, have actively supported ISIS is either ignored or seriously downplayed by both US policy makers and the mainstream media. Do you have any comments on this matter?

The same was true of the Reagan and Bush versions of the "war on terror." For Reagan, it was a pretext to intervene in Central America, in what Salvadoran Bishop Rivera y Damas, who succeeded the assassinated Archbishop Oscar Romero, described as "a war of extermination and genocide against a defenseless civilian population." It was even worse in Guatemala and pretty awful in Honduras. Nicaragua was the one country that had an army to defend it from Reagan's terrorists; in the other countries, the security forces were the terrorists.

In southern Africa, the "war on terror" provided the pretext to support South African crimes at home and in the region, with a horrendous toll. After all, we had to defend civilization from "one of the more notorious terrorist groups" in the world, Nelson Mandela's African National Congress. Mandela himself remained on the US terrorist list until 2008. In the Middle East, the "war on terror" construct led to support for Israel's murderous invasion of Lebanon, and much else. With Bush, it provided a pretext for invading Iraq. And so it continues.

What's happening in the Syrian horror story defies description. The main ground forces opposing ISIS seem to be the Kurds, just as in Iraq, where they are on the US terrorist list. In both countries, they are the prime target of the assault of our NATO ally Turkey, which is also supporting the al-Qaeda affiliate in Syria, al-Nusra Front. The latter seems hardly different from ISIS, though they are having a turf battle. Turkish support for al-Nusra is so extreme that when the Pentagon sent in several dozen fighters it had trained, Turkey apparently alerted al-Nusra, which instantly wiped them out. Al-Nusra and the closely allied Ahrar al-Sham are also supported by US allies Saudi Arabia and Qatar, and, it seems, may be getting advanced weapons from the CIA. It's been reported that they used TOW (theater of war) antitank weapons supplied by the CIA to inflict serious defeats on the Assad army, possibly impelling the Russians to intervene. Turkey seems to be continuing to allow jihadis to flow across the border to ISIS.

Saudi Arabia in particular has been a major supporter of the extremist jihadi movements for years, not only with financing but also by spreading its radical Islamist Wahhabi doctrines with Koranic schools, mosques, and clerics. With no little justice, Middle East correspondent Patrick Cockburn describes the "Wahhabization" of Sunni Islam as one of the most dangerous developments of the era. Saudi Arabia and the Emirates have huge, advanced military forces, but they are barely engaged in the war against ISIS. They do operate in Yemen, where they are creating a major humanitarian catastrophe and very likely, as before, generating future terrorists for us to target in our "war on terror." Meanwhile, the region and its people are being devastated.

For Syria, the only slim hope seems to be negotiations among the many elements involved, excluding ISIS. That includes really awful people, like Syrian president Bashar al-Assad, who are not going to willingly commit suicide and so will have to be involved in negotiations if the spiral to national suicide is not to continue. There are, finally, halting steps in this direction at Vienna. There is more that can be done on the ground, but a shift to diplomacy is essential.

Turkey's role in the so-called global war against terrorism has to be seen as one of the most hypocritical gestures in the modern annals of diplomacy, and Vladimir Putin did not mince his words following the downing of the Russian jet fighter by labeling Turkey "accomplices of terrorists." Oil is the reason why the United States and its Western allies knowingly overlook certain Gulf nations' support for terrorist organizations like ISIS, but what is the reason for neglecting to question Turkey's support of Islamic fundamentalist terrorism?

Turkey has always been an important NATO ally of great geostrategic significance. Through the 1990s, when Turkey was carrying out some of the worst atrocities anywhere in its war against its Kurdish population, it became the leading recipient of US arms (outside Israel and Egypt, a separate category). The relationship has occasionally been under stress, most notably in 2003, when the government adopted the position of 95 percent of the population and refused to join the US attack on Iraq. Turkey was bitterly condemned for this failure to understand the meaning of "democracy." But generally the relationship has remained quite close. Recently, the United

States and Turkey reached an agreement on the war against ISIS: Turkey granted the United States access to the Turkish bases close to Syria and in return pledged to attack ISIS—but instead attacked its Kurdish enemies.

While this may not be a popular view with many people, Russia, unlike the United States, seems to be restrained when it comes to the use of force. Assuming that you agree with this assumption, why do you think this is the case?

They are the weaker party. They don't have eight hundred military bases throughout the world, couldn't possibly intervene everywhere the way the United States has done over the years, or carry out anything like Obama's global assassination campaign. The same was true throughout the Cold War. They could use military force near their borders but couldn't possibly have carried out anything like the Indochina wars, for example.

France seems to have become a favorite target of Islamic fundamentalist terrorists. What's the explanation for that?

Actually, many more Africans are killed by Islamic terrorism. In fact, Boko Haram is ranked higher than ISIS as a global terrorist organization.[1] In Europe, France has been the major target, in large part for reasons going back to the Algerian war.

Islamic fundamentalist terrorism of the kind promoted by ISIS has been condemned by organizations like Hamas and Hezbollah. What differentiates ISIS from other so-called terrorist organizations, and what does ISIS really want?

We have to be careful about what we call "terrorist organizations." Anti-Nazi partisans used terror. So did George Washington's army, so much so that a large part of the population fled in fear of his terror—not to speak of the Indigenous community, for whom he was "the town destroyer." It's hard to find a national liberation movement that hasn't used terror. Hezbollah and Hamas were formed in response to Israeli occupation and aggression. But whatever criteria we use, ISIS is quite different. It is seeking to carve out territory that it will rule and establish an Islamic caliphate. That's quite different from others.

Following the Paris massacre of November 2015, Obama stated in a joint news conference with French President Hollande that "ISIS must be destroyed." Do you think this is possible? If yes, how? If not, why not?

The West does of course have the capacity to slaughter everyone in the ISIS-controlled areas, but even that wouldn't destroy ISIS—or, very likely, some more vicious movement that would develop in its place by the dynamic I mentioned earlier. One goal of ISIS is to draw the "crusaders" into a war with all Muslims. We can contribute to that catastrophe, or we can try to address the roots of the problem and help establish conditions under which the ISIS monstrosity will be overcome by forces within the region.

Foreign intervention has been a curse for a long time, and is likely to continue to be. There are sensible proposals as to how to proceed on this course, for example, the proposal by William Polk, a fine Middle East scholar with rich experience not only in the region but also at the highest levels of US government planning.[2] It receives substantial support from most careful investigations of the appeal of ISIS, notably those of Scott Atran. Unfortunately, the chances that the advice will be heeded are slight.

The political economy of US warfare seems to be structured in such a way that wars appear to be almost inevitable, something which President Dwight Eisenhower was apparently aware of when he warned us in his farewell speech of the dangers of a military-industrial complex. In your view, what will it take to move the United States away from militaristic jingoism?

It is quite true that sectors of the economy benefit from "militaristic jingoism," but I do not think that is its main cause. There are geostrategic and international economic considerations of great import. The economic benefits—only one factor—were discussed in the business press in interesting ways in the early post–World War II period. They understood that massive government spending had rescued the country from the Depression, and there was much concern that if it were curtailed, the country would sink back into depression. One informative discussion, in *Business Week* (February 12, 1949), recognized that social spending could have the same "pump-priming" effect as military spending, but pointed out that for businessmen, "there's a tremendous social and economic difference between welfare pump-priming and military pump-priming." The latter "doesn't really alter the structure of the economy." For the businessman, it's just another order. But welfare and public works spending "does alter the economy. It makes new channels of its own. It creates new institutions. It redistributes income." And we can add more. Military spending scarcely involves the public, but social spending does, and has a democratizing effect. For reasons like these, military spending is much preferred.

Pursuing this question about the link between US political culture and militarism a bit further, is the apparent decline of US supremacy on the global arena more or less likely to turn future US presidents into warmongers?

The United States reached the peak of its power after World War II, but decline set in very soon, first with the "loss of China" and later with the revival of other industrial powers and the agonizing course of decolonization, and in more recent years with other forms of diversification of power. Reactions could take various forms. One is Bush-style triumphalism and aggressiveness. Another is Obama-style reticence to use ground forces. And there are many other possibilities. The popular mood is no slight consideration, and one that we can hope to influence.

Should the left support Bernie Sanders when he caucuses with the Democratic Party?

I think so. His campaign has had a salutary effect. It's raised important issues that are otherwise sidestepped and has moved the Democrats slightly in a progressive direction. Chances that he could be elected in our system of bought elections are not high, and, if he were, it would be extremely difficult for him to effect any significant change of policies. The Republicans won't disappear, and thanks to gerrymandering and other tactics they are likely at least to control the House, as they have done with a minority of votes for some years, and they are likely to have a strong voice in the Senate. The Republicans can be counted on to block even small steps in a progressive—or for that matter even rational—direction. It's important to recognize that they are no longer a normal political party.

As respected political analysts of the conservative American Enterprise

Institute have observed, the former Republican Party is now a "radical insurgency" that has pretty much abandoned parliamentary politics, for interesting reasons that we can't go into here. The Democrats have also moved to the right, and their core elements are not unlike moderate Republicans of years past—though some of Eisenhower's policies would place him about where Sanders is on the political spectrum. Sanders, therefore, would be unlikely to have much congressional support, and would have little at the state level.

Needless to say, the hordes of lobbyists and wealthy donors would hardly be allies. Even Obama's occasional steps in a progressive direction were mostly blocked, though there may be other factors involved, perhaps racism; it's not easy to account for the ferocity of the hatred he has evoked in other terms. But in general, in the unlikely event that Sanders were elected, his hands would be tied—unless, unless, what always matters in the end: unless mass popular movements would develop, creating a wave that he could ride and that might (and should) impel him farther than he might otherwise go.

That brings us, I think, to the most important part of the Sanders candidacy. It has mobilized a huge number of people. If those forces can be sustained beyond the election, instead of fading away once the extravaganza is over, they could become the kind of popular force that the country badly needs if it is to deal in a constructive way with the enormous challenges that lie ahead.

The comments above relate to domestic policies, the areas he has concentrated on. His foreign policy conceptions and proposals seem to me to be pretty conventional liberal Democrat. Nothing particularly novel is proposed, as far as I can see, including some assumptions that I think should be seriously questioned.

One final question. What do you say to those who maintain the view that ending the "war on terror" is naïve and misguided?

Simple: Why? And a more important question: Why do you think that the United States should continue to make major contributions to global terrorism, under the guise of a "war on terror"?

The Empire of Chaos

C. J. POLYCHRONIOU: US military interventions in the twenty-first century (for example, in Afghanistan, Iraq, Libya, Syria) have proven totally disastrous, yet the terms of the intervention debate have yet to be redrawn among Washington's warmakers. What's the explanation for this?

NOAM CHOMSKY: In part, the old cliché—when all you have is a hammer, everything looks like a nail. The comparative advantage of the United States is in military force. When one form of intervention fails, doctrine and practice can be revised with new technologies, devices, and the like. There are possible alternatives, such as supporting democratization (in reality, not rhetoric). But these have likely consequences that the United States would not favor. That is why when the United States supports "democracy"; it is "top-down" forms of democracy, in which traditional elites linked to the United States remain in power, to quote the leading scholar of "democracy promotion," Thomas Carothers, a former Reagan official, who is a strong advocate of the process but who recognizes the reality, unhappily.

Some have argued that Obama's wars are quite different in both style and essence from those of his predecessor, George W. Bush. Is there any validity behind these claims?

Bush relied on shock-and-awe military violence, which proved disastrous for the victims and led to serious defeats for the United States. Obama is

Originally published in *Truthout*, November 5, 2015

relying on different tactics, primarily the drone global assassination campaign, which breaks new records in international terrorism, and Special Forces operations, by now over much of the globe. Nick Turse, the leading researcher on the topic, recently reported that US elite forces were "deployed to a record-shattering 147 countries in 2015."[1]

Destabilization and what I call the "creation of black holes" is the principal aim of the Empire of Chaos in the Middle East and elsewhere, but it is also clear that the United States is sailing in a turbulent sea with no sense of direction and is, in fact, quite clueless in terms of what needs to be done once the task of destruction has been completed. How much of this is due to the decline of the United States as a global hegemon?

The chaos and destabilization are real, but I don't think that's the aim. Rather, it is a consequence of hitting fragile systems that one does not understand with the sledgehammer that is the main tool, as in Iraq, Libya, Afghanistan, and elsewhere. As for the continuing decline of US hegemonic power (actually, from 1945, with some ups and downs), there are consequences in the current world scene. Take, for example, the fate of Edward Snowden. Four Latin American countries are reported to have offered him asylum, no longer fearing the lash of Washington. Not a single European power is willing to face US anger. That is a consequence of very significant decline of US power in the Western Hemisphere.

However, I doubt that the chaos in the Middle East traces substantially to this factor. One consequence of the US invasion of Iraq was to incite sectarian conflicts that are destroying Iraq and are now tearing the region to shreds. The Europe-initiated bombing of Libya created a disaster there, which has spread far beyond with weapons flow and stimulation of jihadi crimes. And there are many other effects of foreign violence. There are also many internal factors. I think that Middle East correspondent Patrick Cockburn is correct in his observation that the "Wahhabization" of Sunni Islam is one of the most dangerous developments of the modern era. By now many of the most horrible problems look virtually insoluble, like the Syrian catastrophe, where the only slim hopes lie in some kind of negotiated settlement toward which the powers involved seem to be slowly inching.

Russia is also raining down destruction in Syria. To what end; and does Russia pose a threat to US interests in the region?

Russian strategy evidently is to sustain the Assad regime, and it is indeed "raining down destruction," primarily attacking the jihadi-led forces supported by Turkey, Saudi Arabia, and Qatar, and to an extent the United States. A recent article in the *Washington Post* suggested that the high-tech weapons provided by the CIA to these forces (including TOW antitank missiles) had shifted the military balance against Assad and were a factor in drawing the Russians in. On "US interest," we have to be careful. The interests of US power and of the people of the United States are often quite different, as is commonly the case elsewhere as well. The official US interest is to eliminate Assad, and naturally Russian support for Assad poses a threat to that. And the confrontation not only is harmful, if not catastrophic, for Syria, but also carries a threat of accidental escalation that could be catastrophic far beyond.

Is ISIS a US-created monster?

A recent interview with the prominent Middle East analyst Graham Fuller is headlined, "Former CIA officer says US policies helped create IS." What Fuller says, correctly I think, is that

> I think the United States is one of the key creators of this organization. The United States did not plan the formation of ISIS, but its destructive interventions in the Middle East and the war in Iraq were the basic causes of the birth of ISIS. You will remember that that the starting point of this organization was to protest the US invasion of Iraq. In those days it was supported by many non-Islamist Sunnis as well because of their opposition to Iraq's occupation. I think even today ISIS [now the Islamic State] is supported by many Sunnis who feel isolated by the Shiite government in Baghdad.

Establishment of Shiite dominance was one direct consequence of the US invasion, a victory for Iran and one element of the remarkable US defeat in Iraq. So in answer to your question, US aggression was a factor in the rise of ISIS, but there is no merit to conspiracy theories circulating in the region that hold that the United States planned the rise of this extraordinary monstrosity.

How do you explain the fascination that a completely barbaric and savage organization like the Islamic State holds for many young Muslim people living in Europe?

There has been a good deal of careful study of the phenomenon, by Scott Atran among others. The appeal seems to be primarily among young people who live under conditions of repression and humiliation, with little hope and little opportunity, and who seek some goal in life that offers dignity and self-realization; in this case, establishing a utopian Islamic state rising in opposition to centuries of subjugation and destruction by Western imperial power. In addition, there appears to be a good deal of peer pressure—members of the same soccer club, and so on. The sharply sectarian nature of the regional conflicts no doubt is also a factor—not just "defending Islam" but defending it from Shiite apostates. It's a very ugly and dangerous scene.

The Obama administration has shown little interest in reevaluating the US relationship with authoritarian and fundamentalist regimes in places like Egypt and Saudi Arabia. Is democracy promotion a completely sham element of US foreign policy?

There doubtless are people like Thomas Carothers, mentioned above, who really are dedicated to democracy promotion, and are within the government; he was involved in "democracy promotion" in the Reagan State Department. But the record shows quite clearly that it is scarcely an element in policy, and quite often democracy is considered a threat—for good reasons, when we look at popular opinion. To mention only one obvious example, polls of international opinion by the leading US polling agency (WIN/Gallup) show that the United States is regarded as the greatest threat to world peace by a huge margin, Pakistan far behind in second place (presumably inflated by the Indian vote). Polls taken in Egypt on the eve of the Arab Spring revealed considerable support for Iranian nuclear weapons to counterbalance Israeli and US power. Public opinion often favors social reform of the kind that would harm US-based multinationals. And much else. These are hardly policies that the US government would like to see instituted, but authentic democracy would give a significant voice to public opinion. For similar reasons, democracy is feared at home.

Do you anticipate any major changes in US foreign policy in the near future, either under a Democratic or Republican administration?

Not under a Democratic administration, but the situation with a Republican administration is much less clear. The party has drifted far off the spectrum of parliamentary politics. If the pronouncements of the current crop of candidates can be taken seriously, the world could be facing deep trouble. Take, for example, the nuclear deal with Iran. Not only are they unanimously opposed to it, but they are competing on how quickly to bomb Iran. It's a very strange moment in American political history, and in a state with awesome powers of destruction that should cause not a little concern.

Global Struggles for Dominance: ISIS, NATO, and Russia

C. J. POLYCHRONIOU: The rise of ISIS is a direct consequence of the US invasion and occupation of Iraq and represents today, by far, the most brutal and dangerous terrorist organization we have seen in recent memory. It also appears that its tentacles have reached beyond the "black holes" created by the United States in Syria, Libya, Iraq, and Afghanistan and have now taken hold inside Europe, a fact acknowledged recently by German chancellor Angela Merkel. In fact, it has been estimated that attacks organized or inspired by ISIS have taken place every forty-eight hours in cities outside the above-mentioned countries since early June 2016. Why have countries like Germany and France become the targets of ISIS?

NOAM CHOMSKY: I think we have to be cautious in interpreting ISIS claims of responsibility for terrorist attacks. Take the worst of the recent ones, in Nice. It was discussed by Akbar Ahmed, one of the most careful and discerning analysts of radical Islam. He concludes from the available evidence that the perpetrator, Mohamed Lahouaiej Bouhlel, was probably "not a devout Muslim. He had a criminal record, drank alcohol, ate pork, did drugs, did not fast, pray or regularly attend a mosque and was not religious in any way. He was cruel to his wife, who left him. This is not what many Muslims would typically consider reflective of their faith, particularly those who consider themselves religiously devout." ISIS did (belatedly) "take credit" for the

---

Originally published in *Truthout*, August 17, 2016

attack, as they routinely do, whatever the facts, but Ahmed regards the claim as highly dubious in this case. On this and similar attacks, he concludes that

> the reality is that while ISIS may influence these Muslims in a general way, their animus is coming from their position as unwanted immigrants in Europe, especially in France, where they are still not treated [as] French, even if they are born there. The community as a whole has a disproportionate population of unemployed youth with poor education and housing and is constantly the butt of cultural humiliation. It is not an integrated community, barring some honorable exceptions. From it come the young men like Lahouaiej Bouhlel. The pattern of [the] petty criminal may be observed in the other recent terrorist attacks in Europe, including those in Paris and Brussels.

Ahmed's analysis corresponds closely to that of others who have done extensive investigation of recruits to ISIS, notably Scott Atran and his research team. And it should, I think, be taken seriously, along with his prescriptions, which also are close to those of other knowledgeable analysts: to "provide the Muslim community educational and employment opportunities, youth programs, and promote acceptance, diversity and understanding. There is much that governments can do to provide language, cultural and religious training for the community, which will help resolve, for example, the problem of foreign imams having difficulty transferring their roles of leadership into local society."

Merely to take one illustration of the problem to be faced, Atran points out that "only 7 to 8 percent of France's population is Muslim, whereas 60 to 70 percent of France's prison population is Muslim." It's also worth taking note of a recent National Research Council report, which found that "with respect to political context, terrorism and its supporting audiences appear to be fostered by policies of extreme political repression and discouraged by policies of incorporating both dissident and moderate groups responsibly into civil society and the political process."

It's easy to say, "Let's strike back with violence"—police repression, carpet-bomb them to oblivion (Ted Cruz), and so on—very much what al-Qaeda and ISIS have hoped for, and very likely to intensify the problems, as, indeed, has been happening until now.

What is ISIS's aim when targeting innocent civilians, such as the attack on the seaside town of Nice in France in which eighty-four people were killed?

As I mentioned, we should, I think, be cautious about the claims and charges of ISIS initiative, or even involvement. But when they are involved in such atrocities, the strategy is clear enough. Careful and expert analysts of ISIS and violent insurgencies (Scott Atran, William Polk, and others) generally tend to take ISIS at its word. Sometimes they cite the "playbook" in which the core strategy used by ISIS is laid out, written a decade ago by the Mesopotamian wing of the al-Qaeda affiliate that morphed into ISIS. Here are the first two axioms (quoting an article by Atran):

> [Axiom 1:] Hit soft targets: "Diversify and widen the vexation strikes against the Crusader-Zionist enemy in every place in the Islamic world, and even outside of it if possible, so as to disperse the efforts of the alliance of the enemy and thus drain it to the greatest extent possible." [Axiom 2:] Strike when potential victims have their guard down to maximise fear in general populations and drain their economies: "If a tourist resort that the Crusaders patronise . . . is hit, all of the tourist resorts in all of the states of the world will have to be secured by the work of additional forces, which are double the ordinary amount, and a huge increase in spending."

And the strategy has been quite successful, both in spreading terrorism and imposing great costs on the "Crusaders" with slight expenditure.

It has been reported that tourists in France will be protected by armed forces and soldiers at holiday sites, including beaches. How much of this development is linked to the refugee crisis in Europe, where millions have been arriving in the last couple of years from war-torn regions around the world?

Hard to judge. The crimes in France have not been traced to recent refugees, as far as I have seen. Rather, it seems to be more like the Lahouaiej Bouhlel case. But there is great fear of refugees, far beyond any evidence relating them to crime. Much the same appears to be true in the United States, where Trump-style rhetoric about Mexico sending criminals and rapists doubtless frightens people, even though the limited statistical evidence indicates that "first-generation immigrants are predisposed to lower crime rates than native-born Americans," as reported by Michelle Ye Hee Lee in the *Washington Post*.

To what extent would you say that Brexit was being driven by xenophobia and the massive inflow of immigrants into Europe?

There has been plenty of reporting giving that impression, but I haven't seen any hard data. And it's worth recalling that the inflow of immigrants is from the EU, not those fleeing from conflict. It's also worth recalling that Britain has had a nontrivial role in generating refugees. The invasion of Iraq, to give one example. Many others, if we consider greater historical depth. The burden of dealing with the consequences of US-UK crimes falls mainly on countries that had no responsibility for them, like Lebanon, where about 40 percent of the population is estimated to be refugees.

Are the United States and the major Western powers really involved in a war against ISIS? This would seem doubtful to an outside observer, given the growing influence of ISIS and the continuing ability of the organization to recruit soldiers for its cause from inside Europe.

Speculations to that effect are rampant in the Middle East, but I don't think they have any credibility. The United States is powerful, but not all-powerful. There is a tendency to attribute everything that happens in the world to the CIA or some diabolical Western plan. There is plenty to condemn, sharply. And the United States is indeed powerful. But it's nothing like what is often believed.

There seems to be a geopolitical shift underway in Turkey's regional political role, which may have been the ultimate cause behind the failed coup of July 2016. Do you detect such a shift under way?

There certainly has been a shift in regional policy from former Turkish Prime Minister Davutoğlu's "Zero Problems Policy," but that's because problems abound. The goal of becoming a regional power, sometimes described as neo-Ottoman, seems to be continuing, if not accelerating. Relations with the West are becoming more tense as Erdoğan's government continues its strong drift toward authoritarian rule, with quite extreme repressive measures. That naturally impels Turkey to seek alliances elsewhere, particularly with Russia. Erdoğan's first post-coup visit was to Moscow, in order to restore "the Moscow-Ankara friendship axis" (in his words) to what it was before Turkey shot down a Russian jet in November 2015 when it allegedly passed across the Turkish border for a few seconds while on a bombing mission in Syria. Very unfortunately, there is very little Western opposition to Erdoğan's violent and vicious escalation of atrocities against the Kurdish population in the southeast, which some observers now describe as approaching the horrors of the 1990s. As for the coup, its background remains obscure, for the time being. I don't know of evidence that shifts in regional policy played a role.

The coup against Erdoğan ensured the consolidation of a highly authoritarian regime in Turkey: Erdoğan arrested thousands of people and closed down media outlets, schools, and universities following the coup. The effects of the coup may, in fact, even strengthen the role of the military in political affairs as it will come under the direct control of the president himself, a move that Erdoğan has already initiated. How will this affect Turkey's relations with the United States and European powers, given the alleged concerns of the latter about human rights and democracy inside Turkey and about Erdoğan's pursuit of closer ties with Putin?

The correct word is "alleged." During the 1990s, the Turkish government was carrying out horrifying atrocities, targeting its Kurdish population—tens of thousands killed, thousands of villages and towns destroyed, hundreds of thousands (maybe millions) driven from their homes, every imaginable form of torture. Eighty percent of the arms were coming from Washington, increasing as atrocities increased. In the single year 1997, when atrocities were peaking, Clinton sent more arms than the sum total sent to Turkey throughout the entire postwar era until the onset of the counterinsurgency campaign. The media virtually ignored all of this. The *New York Times* has a bureau in Ankara, but it reported almost nothing. The facts were, of course, widely known in Turkey—and elsewhere, to those who took the trouble to look. Now that atrocities are peaking again, as I mentioned, the West prefers to look elsewhere.

Nevertheless, relations between Erdoğan's regime and the West are becoming more tense, and there is great anger against the West among Erdoğan supporters because of Western attitudes toward the coup (mildly critical, but not enough for the regime) and toward the increased authoritarianism and sharp repression (mild criticism, but too much for the regime). In fact, it is widely believed that the United States initiated the coup.

The United States is also condemned for asking for evidence before extraditing Gulen, whom Erdoğan blames for the coup. Not a little irony here. One may recall that the United States bombed Afghanistan because the Taliban refused to turn Osama bin Laden over without evidence. Or take the case of Emmanuel "Toto" Constant, the leader of the terrorist force FRAPH (Front for the Advancement and Progress of Haiti) that ran wild in Haiti under the military dictatorship of the early '90s. When the junta was overthrown by a Marine invasion, he escaped to New York, where he was living comfortably. Haiti wanted him extradited and had more than enough evidence. But Clinton refused, very likely because he would have exposed Clinton's ties to the murderous military junta.

The recent migration deal between Turkey and the EU seems to be falling apart, with Erdoğan having gone so far as to say publicly that "European leaders are not being honest." What could be the consequences for Turkey–EU relations, and for the refugees themselves, if the deal were to fall apart?

Basically, Europe bribed Turkey to keep the miserable refugees—many fleeing from crimes for which the West bears no slight responsibility—from reaching Europe. It is similar to Obama's efforts to enlist Mexican support in keeping Central American refugees—often very definitely victims of US policies, including those of the Obama administration—from reaching the US border. Morally grotesque, but better than letting them drown in the Mediterranean. The deterioration of relations will probably make their travail even worse.

NATO, still a US-dominated military alliance, has increased its presence in Eastern Europe lately, as it is bent on stopping Russia's revival by creating divisions between Europe and Russia. Is the United States looking for a military conflict with Russia, or are such moves driven by the need to keep the military-industrial complex intact in a post–Cold War world?

NATO is surely a US-dominated military alliance. As the USSR collapsed, Russia's Mikhail Gorbachev proposed a continent-wide security system, which the United States rejected, insisting on preserving NATO—and expanding it. Gorbachev agreed to allow a unified Germany to join NATO, a remarkable concession in the light of history. There was, however, a quid pro quo: that NATO not expand "one inch to the east," meaning to East Germany. That was promised by President Bush I and secretary of state James Baker, but not on paper; it was a verbal commitment, and the United States later claimed that means it was not binding.

Careful archival research by Joshua R. Itzkowitz Shifrinson, published last spring in the prestigious Harvard-MIT journal *International Security*, reveals very plausibly that this was intentional deceit, a very significant discovery that substantially resolves, I think, scholarly dispute about the matter. NATO did expand to East Germany; in later years, to the Russian border. Those plans were sharply condemned by George Kennan and other highly respected commentators because they were very likely to lead to a new Cold War, as Russia naturally felt threatened. The threat became more severe when NATO invited Ukraine to join in 2008 and 2013. As Western analysts recognize, that extends the threat to the core of Russian strategic concerns, a matter discussed, for example, by John Mearsheimer in the lead article in the major establishment journal *Foreign Affairs*.

However, I do not think the goal is to stop Russia's revival or to keep the military-industrial complex intact. And the United States certainly doesn't want a military conflict, which would destroy both sides (and the world). Rather, I think it's the normal effort of a great power to extend its global dominance. But it does increase the threat of war, if only by accident, as Kennan and others presciently warned.

In your view, does a nuclear war between the United States and Russia remain a very real possibility in today's world?

A very real possibility, and in fact, an increasing one. That's not just my judgment. It's also the judgment of the experts who set the Doomsday Clock of the *Bulletin of Atomic Scientists*; of former defense secretary William Perry, one of the most experienced and respected experts on these matters; and of numerous others who are by no means scaremongers. The record of near accidents, which could have been terminal, is shocking, not to speak of very dangerous adventurism. It is almost miraculous that we have survived the nuclear weapons era, and playing with fire is irresponsible in the extreme. In fact, these weapons should be removed from the Earth, as even many of the most conservative analysts recognize—Henry Kissinger, George Shultz, and others.

Is European Integration Unraveling?

C. J. POLYCHRONIOU: Noam, thanks for doing this interview on current developments in Europe. I would like to start by asking you this question: Why do you think Europe's refugee crisis is happening now?

NOAM CHOMSKY: The crisis has been building up for a long time. It is hitting Europe now because it has burst the bounds, from the Middle East and from Africa. Two Western sledgehammer blows had a dramatic effect. The first was the US-UK invasion of Iraq, which dealt a nearly lethal blow to a country that had already been devastated by a massive military attack twenty years earlier, followed by virtually genocidal US-UK sanctions. Apart from the slaughter and destruction, the brutal occupation ignited a sectarian conflict that is now tearing the country and the entire region apart. The invasion displaced millions of people, many of whom fled and were absorbed in the neighboring countries, poor countries that are left to deal somehow with the detritus of our crimes.

One outgrowth of the invasion is the ISIS/Daesh monstrosity, which is contributing to the horrifying Syrian catastrophe. Again, the neighboring countries have been absorbing the flow of refugees. Turkey alone has over 2 million Syrian refugees. At the same time it is contributing to the flow by its policies in Syria: supporting the extremist al-Nusra Front and other radical Islamists and attacking the Kurds who are the main ground force opposing

Originally published in *Truthout*, January 25, 2016

ISIS—which has also benefited from not-so-tacit Turkish support. But the flood can no longer be contained within the region.

The second sledgehammer blow destroyed Libya, now a chaos of warring groups, an ISIS base, a rich source of jihadis and weapons from West Africa to the Middle East, and a funnel for the flow of refugees from Africa. That at once brings up longer-term factors. For centuries, Europe has been torturing Africa—or, to put it more mildly—exploiting Africa for Europe's own development, to adopt the recommendation of the top US planner, George Kennan, after World War II.

The history, which should be familiar, is beyond grotesque. To take just a single case, consider Belgium, now groaning under a refugee crisis. Its wealth derived in no small measure from "exploiting" the Congo with brutality that exceeded even that of its European competitors. Congo finally won its freedom in 1960. It could have become a rich and advanced country once freed from Belgium's clutches, spurring Africa's development as well. There were real prospects, under the leadership of Patrice Lumumba, one of the most promising figures in Africa. He was targeted for assassination by the CIA, but the Belgians got there first. His body was cut to pieces and dissolved in sulfuric acid. The United States and its allies supported the murderous kleptomaniac Mobutu. By now Eastern Congo is the scene of the world's worst slaughters, assisted by US favorite Rwanda, while warring militias feed the craving of Western multinationals for minerals for cell phones and other high-tech wonders. The picture generalizes too much of Africa, exacerbated by innumerable crimes. For Europe, all of this becomes a refugee crisis.

Do the waves of immigrants (obviously many of them are immigrants, not simply refugees from war-torn regions) penetrating the heart of Europe represent some kind of a "natural disaster," or is it purely the result of politics?

There is an element of natural disaster. The terrible drought in Syria that shattered the society was presumably the effect of global warming, which is not exactly natural. The Darfur crisis was in part the result of desertification that drove nomadic populations to settled areas. The awful Central African famines today may also be in part due to the assault on the environment during the Anthropocene, the new geological era when human activities, mainly industrialization, have been destroying the prospects for decent survival, and will do so, unless curbed.

European Union officials are having an exceedingly difficult time coping with the refugee crisis because many EU member states are unwilling to do their part and accept anything more than just a handful of refugees. What does this say about EU governance and the values of many European societies?

EU governance works very efficiently to impose harsh austerity measures that devastate poorer countries and benefit Northern banks. But it has broken down almost completely when addressing a human catastrophe that is in substantial part the result of Western crimes. The burden has fallen on the few who were willing, at least temporarily, to do more than lift a finger, like Sweden and Germany. Many others have just closed their borders. Europe is trying to induce Turkey to keep the miserable wrecks away from its borders, just as the United States is doing, pressuring Mexico to prevent those trying to escape the ruins of US crimes in Central America from reaching US borders. This is even described as a humane policy that reduces "illegal immigration."

What does all of this tell us about prevailing values? It is hard even to use the word "values," let alone to comment. That's particularly when writing in the United States, probably the safest country in the world, now consumed by a debate over whether to allow Syrians in at all because one might be a terrorist pretending to be a doctor, or, at the extremes, which unfortunately is in the US mainstream, whether to allow any Muslims in at all, while a huge wall protects us from immigrants fleeing from the wreckage south of the border.

What about the argument that it is simply impossible for many European countries to accommodate so many immigrants and refugees?

Germany has done the most, absorbing about 1 million refugees in a very rich country of over 80 million people. Compare Lebanon, a poor country with severe internal problems. Its population is now about 25 percent Syrian, in addition to the descendants of those who were expelled from the former Palestine. Furthermore, unlike Lebanon, Germany badly needs immigrants to maintain its population with the declining fertility that has tended to result from education of women, worldwide. Kenneth Roth, the head of Human Rights Watch, is surely right to observe that "this 'wave of people' is more like a trickle when considered against the pool that must absorb it. Considering the EU's wealth and advanced economy, it is hard to argue that Europe lacks the means to absorb these newcomers," particularly in countries that need immigrants for their economic health.

Many of the refugees trying to get to Europe never make the journey, with many dead washing up on Greece's and Italy's shores. In fact, according to the UN refugee agency, the UN High Commissioner for Refugees (UNHCR), more than 2,500 people have died this past summer [2015] alone trying to cross the Mediterranean to Europe, with the southwestern coast of Turkey having become the departure point for thousands of refugees who are lured into crumbling boats by Turkish migrant smugglers. Why isn't Europe putting more pressure on the Turkish government of president Recep Tayyip Erdoğan to do something about this horrible situation?

The primary European efforts, as noted, have been to pressure Turkey to keep the misery and suffering far from us. Much like the United States and Mexico. Their fate, once we are safe from the contagion, is of much lesser concern.

Just recently, you accused Erdoğan of double standards on terrorism when he singled you out for a petition signed by hundreds of academicians protesting Turkey's actions against the Kurdish population, calling you, in fact, a terrorist. Can you say a few things about this matter, since it evolved into an international incident?

It is fairly straightforward. A group of Turkish academics initiated a petition protesting the government's severe and mounting repression of its Kurdish population. I was one of several foreigners invited to sign. Immediately after a murderous terrorist attack in Istanbul, Erdoğan launched a tirade bitterly attacking the signers of the declaration, declaring Bush-style that you are either with us or with the terrorists. Since he singled me out for a stream of invective, I was asked by Turkish media and friends to respond. I did so, briefly, as follows: "Turkey blamed ISIS, which Erdoğan has been aiding in many ways, while also supporting the al-Nusra Front, which is hardly different. He then launched a tirade against those who condemn his crimes against Kurds—who happen to be the main ground force opposing ISIS in both Syria and Iraq. Is there any need for further comment?"

Turkish academics who signed the petition were detained and threatened; others were physically attacked. Meanwhile state repression continues to escalate. The dark days of the 1990s have hardly faded from memory. As before, Turkish academics and others have demonstrated remarkable courage and integrity in vigorously opposing crimes of state, in a manner rarely to be found elsewhere, risking and sometimes enduring severe punishment for their honorable stance. There is, fortunately, growing international support for them, though it still falls far short of what is merited.

In a correspondence we had, you referred to Erdoğan as "the dictator of his dreams." What do you mean by this?

For several years, Erdoğan has been taking steps to consolidate his power, reversing the encouraging steps toward democracy and freedom in Turkey in earlier years. He shows every sign of seeking to become an extreme authoritarian ruler, approaching dictatorship, and a harsh and repressive one.

The Greek crisis continues unabated, and the country's international creditors are demanding constantly additional reforms of the kind that no democratic government anywhere else in Europe would be able to implement. In some cases, in fact, their demands for more reforms are not accompanied by specific measures, giving one the impression that what is going on is nothing more than a display of brutal sadism toward the Greek people. What are your views on this matter?

The conditions imposed on Greece in the interests of creditors have devastated the country. The proclaimed goal was to reduce the debt burden, which has increased under these measures. As the economy has been undermined, GDP has naturally declined, and the debt-to-GDP ratio has increased despite radical slashing of state expenditures. Greece has been provided with debt relief, theoretically. In reality, it has become a funnel through which European aid flows to the Northern banks that made risky loans that failed and want to be bailed out by European taxpayers, a familiar feature of financial institutions in the neoliberal age.

When the Greek government suggested asking the people of Greece to express their opinions on their fate, the reaction of European elites was utter horror at the impudence. How can Greeks dare to regard democracy as a value to be respected in the country of its origin? The ruling Eurocrats reacted with utter sadism, imposing even harsher demands to reduce Greece to ruins, meanwhile, no doubt, appropriating what they can for themselves. The target of the sadism is not the Greek people specifically, but anyone who dares to imagine that people might have rights that begin to compare with those of financial institutions and investors. Quite generally, the measures of austerity during recession made no economic sense, as recognized even by the economists of the IMF (though not its political actors). It is difficult to regard them as anything other than class war, seeking to undo the social democratic gains that have been one of Europe's major contributions to modern civilization.

And your views on the Syriza-led government, which has reneged on its pre-election promises and ended up signing a new bailout agreement, thereby becoming yet another Greek government enforcing austerity and antipopular measures?

I do not feel close enough to the situation to comment on Syriza's specific choices or to evaluate alternative paths that it might have pursued. Their options would have been considerably enhanced had they received meaningful support from popular forces elsewhere in Europe, as I think could have been possible.

The former Greek finance minister, Yanis Varoufakis, is about to launch a new party whose aim is to carry out, as he said, "a simple but radical idea: to democratize Europe." I have two questions for you on this matter: First, why is social democracy becoming increasingly a thing of the

past in many European societies? And, second how far can one "democratize" capitalism?

Social democracy, not just its European variant but others as well, has been under severe attack through the neoliberal period of the past generation, which has been harmful to the general population almost everywhere while benefiting tiny elites. One illustration of the obscenity of these doctrines is revealed in the study, just released by Oxfam, finding that the richest 1 percent of the world's population will soon hold more than half of the world's wealth. Meanwhile, in the United States, the richest of the world's major societies and with incomparable advantages, millions of children live in households that try to survive on two dollars a day. Even that pittance is under attack by so-called conservatives.

How far reforms can proceed under the existing varieties of state capitalism, one can debate. But that they can go far beyond what now exists is not at all in doubt. Nor is it in doubt that every effort should be made to press them to their limits. That should be a goal even for those committed to radical social revolution, which would only lead to worse horrors if it were not to arise from the dedication of the great mass of the population who come to realize that that the centers of power will block further steps forward.

Europe's refugee crisis has forced several EU member states, including Austria, Sweden, Denmark, and the Netherlands, to suspend the Schengen Agreement. Do you think we are in the midst of witnessing the unraveling of the EU integration project, including perhaps the single currency?

I think we should distinguish between the single currency, for which circumstances were not appropriate, and the EU integration project, which, I think, has been a major advance. It is enough to recall that for hundreds of years Europe was devoted to mutual slaughter on a horrific scale. Overcoming of national hostilities and erosion of borders is a substantial achievement. It would be a great shame if the Schengen Agreement collapses under a perceived threat that should not be difficult to manage in a humane way, and might indeed contribute to the economic and cultural health of European society.

Burkini Bans, New Atheism, and State Worship: Religion in Politics*

C. J. POLYCHRONIOU: In the course of human history, religion has provided relief from pain and suffering to poor and oppressed people around the world, which is probably what Marx meant when he said, "Religion is the opium of the people." But, at the same time, unspeakable atrocities have been committed in the name of God, and religious institutions often function as the guardians of tradition. What are your own views on the role of religion in human affairs?

NOAM CHOMSKY: The general picture is quite ugly and too familiar to recount. But it is worth remembering that there are some exceptions. One striking example is what happened in Latin America after Vatican II in 1962, called at the initiative of Pope John XXIII. The proceedings took significant steps toward restoring the radical pacifist message of the Gospels that had been largely abandoned when the Emperor Constantine, in the fourth century, adopted Christianity as the official doctrine of the Roman Empire—turning the church of the persecuted into the church of the persecutors, as historian of Christianity Hans Küng described the transformation. The message of Vatican II was taken up in Latin America by bishops, priests, lay persons who devoted themselves to helping poor and bitterly oppressed people to organize to gain and defend their rights—what came to be called "liberation theology."

*Coauthored with Lily Sage; originally published in *Truthout*, August 31, 2016

There were, of course, earlier roots and counterparts in many Protestant denominations, including evangelical Christians. These groups formed a core part of a remarkable development in the United States in the 1980s when, for the first time ever to my knowledge, a great many people not only protested the terrible crimes that their government was committing but went to join and help the victims to survive the onslaught.

The United States launched a virtual war against the church, most dramatically in Central America in the 1980s. The decade was framed by two crucial events in El Salvador: the assassination in 1980 of Archbishop Oscar Romero, the "voice for the voiceless," and the assassination of six leading Latin American intellectuals, Jesuit priests, in 1989. Romero was assassinated a few days after he sent an eloquent letter to President Carter pleading with him not to send aid to the murderous military junta, who would use it "to destroy the people's organizations fighting to defend their fundamental human rights," in Romero's words. So the security forces did, in the US-dominated states of the region, leaving many religious martyrs along with tens of thousands of the usual victims: poor peasants, human rights activists, and others seeking "to defend their fundamental human rights."

The US military takes pride in helping to destroy the dangerous heresy that adopted "the preferential option for the poor," the message of the Gospels. The School of the Americas (renamed The Western Hemisphere Institute for Security Cooperation), famous for training of Latin American killers, announces proudly that liberation theology was "defeated with the assistance of the US army."

Do you believe in the spiritual factor behind religion or find something useful in it?

For me, personally, no. I think irrational belief is a dangerous phenomenon and I try to avoid it. On the other hand, I recognize that it's a significant part of the lives of others, with mixed effects.

What are your views on the rise of "new atheism," which seems to have come about in response to the 9/11 terrorist attacks? Who are this

movement's target audiences, and does it have a distinguishable political agenda around which the progressive and left forces should rally?

It's often not very clear who the target audiences are, and agendas no doubt vary. It's fine to carry out educational initiatives aimed at encouraging people to question baseless and irrational beliefs, which can often be quite dangerous. And perhaps, sometimes such efforts have positive effects. But questions arise.

Take, for example, George W. Bush, who invoked his fundamentalist Christian beliefs in justifying his invasion of Iraq, the worst crime of the century. Is he part of the intended audience, or his variety of evangelical Christians? Or the prominent rabbis in Israel who call for visiting the judgment of Amalek on all Palestinians (total destruction, down to their animals)? Or the radical Islamic fundamentalists in Saudi Arabia who have been Washington's highly valued allies in the Middle East for seventy-five years, while they have been implementing the Wahhabization of Sunni Islam? If groups like these are the intended audiences of "new atheism," the effort is not very promising, to say the least. Is it people with no particular religious beliefs who attend religious ceremonies regularly and celebrate holidays so that they can become part of a community of mutual support and solidarity, and together with others enjoy a tradition and reinforce values that help overcome the isolation of an atomized world lacking social bonds? Is it the grieving mother who consoles herself by thinking that she will see her dying child again in heaven? No one would deliver solemn lectures on epistemology to her. There may indeed be an audience, but its composition and bounds raise questions.

Furthermore, to be serious, the "new atheism" should target the virulent secular religions of state worship, often disguised in the rhetoric of exceptionalism and noble intent, the source of crimes so frequent and immense that recounting them is hardly necessary.

Without going on, I have reservations. Though, again, efforts to overcome false and often extremely dangerous beliefs are always appropriate.

One could make the argument that the United States is in reality a deeply fundamentalist country when it comes to the issue of religion. Is there a hope for true progressive change in this country when the overwhelming bulk of the population seems to be in the grip of religious fervor?

The United States has been a deeply fundamentalist country since its origins, with repeated Great Awakenings and outbursts of religious fervor. It stands out today among the industrial societies in the power of religion. Nevertheless, also from its origins there has been significant progressive change, and it has not necessarily been in conflict with religious commitments.

One thinks, for example, of Dorothy Day and the Catholic Worker movement. Or of the powerful role of religion in African American communities in the great civil rights movement—and as a personal aside, it was deeply moving to be able to take part in meetings of demonstrators in churches in the South after a day of brutal beatings and savagery, where the participants were reinforcing bonds of solidarity, singing hymns, gathering strength to go on the next day. This is, of course, by no means the norm, and commonly the impact of fundamentalist religious commitment on social policy has been harmful, if not pernicious.

As usual, there are no simple answers, just the old familiar ones: sympathetic concern, efforts to bring out what is constructive and worthwhile and to overcome harmful tendencies, and to continue to develop the forces of secular humanism and far-reaching and radical commitments that are urgently needed to deal with the pressing and urgent problems we all face.

So many political speeches in the United States end with, "God bless you, and God bless America." Do linguistic expressions like these influence politics, culture, and social reality?

I presume the causal relation is substantially in the opposite direction, though there may well be feedback. A drumbeat of propaganda on how "we are good" and "they are evil," with constant exercises of self-admiration and abuse of others, can hardly fail to have an impact on perception of the world.

Examples abound, but merely to illustrate the common pattern, take a current example from the peak of the intellectual culture: Samantha Power's August 18, 2016, article in the *New York Review of Books*. Without any relevant qualification or comment, the author presents Henry Kissinger's sage reflections on "America's tragic flaw": namely, "believing that our principles are universal principles, and seeking to extend human rights far beyond our nation's borders . . . . 'No nation . . . has ever imposed the moral demands on itself that America has. And no country has so tormented itself over the gap between its moral values, which are by definition absolute, and the imperfection inherent in the concrete situations to which they must be applied.'"

For anyone with the slightest familiarity with contemporary history, such fatuous musings are simply an embarrassment—or to be more accurate, a horror. And this is not talk radio, but a leading journal of left-liberal intellectuals. People bombarded with patriotic drivel from all corners are likely to have a view of themselves and the world that poses major threats to humanity.

Rhetoric is widely used in political campaigns and is frequently abused in a political context. Do you have a theory of political rhetoric?

I don't have any theory of rhetoric, but I try to keep in mind the principle that one should not try to persuade; rather, one should lay out the territory as best one can so that others can use their own intellectual powers to determine for themselves what they think is taking place and what is right or wrong. I also try, particularly in political writing, to make it extremely clear in advance exactly where I stand so that readers can make judgments accordingly. The idea of neutral objectivity is at best misleading and often fraudulent. We cannot help but approach complex and controversial questions—especially those of human significance—with a definite point of view, with an ax to grind if you like, and that ax should be apparent right up front so that those we address can see where we are coming from in our choice and interpretation of the events of history.

To the extent that I can monitor my own rhetorical activities, which is probably not a lot, I try to refrain from efforts to bring people to reach my conclusions without thinking the matter through on their own. Similarly, any good teacher knows that conveying information is of far less importance than helping students gain the ability to inquire and create on their own.

It has become popular over the years to think of knowledge as something that is socially constructed, and proponents of the idea that knowledge is simply the outcome of a consensus on any subject matter requiring research and analysis say the same goes for reality itself. Do you agree with this relativistic view of knowledge and reality?

I think it is mostly far off track, though there is an element of truth hidden within. No doubt the pursuit of knowledge is guided by prior conceptions, and no doubt it is often, not always, but typically, a communal activity. That's substantially true of organized knowledge, say research in the natural sciences. For example, a graduate student will come in and inform me I was wrong about what I said in a lecture yesterday for this or that reason, and we'll discuss it, and we'll agree or disagree, and maybe another set of problems will come out. Well, that's normal inquiry, and whatever results is some form of knowledge or understanding, which is, in part, socially determined by the nature of these interactions.

There is a great deal that we don't understand much about, like how scientific knowledge is acquired and develops. If we look more deeply at the domains where we do understand something, we discover that the development of cognitive systems, including systems of knowledge and understanding, is substantially directed by our biological nature. In the case of knowledge of language, we have clear evidence and substantial results about this. Part of my own personal interest in the study of language is that it's a domain in which these questions can be studied fairly clearly, much more so than in many others. Also, it's a domain that is intrinsic to human nature and human functions, not a marginal case. Here, I think, we have very powerful evidence of the directive effect of biological nature on the form of the system of knowledge that arises.

In other domains like, for example, the internal construction of our moral code, we just know less, though there is quite interesting and revealing current research into the topic. I think the qualitative nature of the problem faced strongly suggests a very similar conclusion: a highly directive effect of biological nature. When you turn to scientific inquiry, again, so little is known about how it proceeds—how discoveries are made—that we are reduced to speculation and review of historical examples. But I think the qualitative nature of the process of acquiring scientific knowledge again suggests a highly directive effect of biological nature. The reasoning behind this is basically Plato's, which I think is essentially valid. That's why it's sometimes called "Plato's problem." The reasoning in the Platonic dialogues is that the richness and specificity and commonality of the knowledge we attain is far beyond anything that can be accounted for by the experience available, which includes interpersonal interactions. And, apart from acts of God, that leaves only the possibility that it's inner-determined in essential ways, ultimately by biological endowment.

That's the same logic that's routinely used by natural scientists studying organic systems. So, for example, when we study physical growth—metaphorically speaking, "below the neck," everything but the mind—we take this reasoning for granted. Let's say I were to suggest to you that undergoing puberty is a matter of social interaction and people do it because they see other people do it, that it's peer pressure. Well, you would laugh. Why? There is nothing in the environment that could direct these highly specific changes in the organism. Accordingly, we all take for granted that it is biologically determined, that growing children are somehow programmed to undergo puberty at a certain stage of development. Are social factors irrelevant to puberty? No, not at all. Social interaction is certainly going to be relevant. Under certain conditions of social isolation, it might not even take place. The same logic holds when inquiry proceeds "above the neck."

Returning to the subject of the link between religion and politics, it has been argued by quite a few commentators that the Israeli–Palestinian conflict is a war of religion, not territory. Any validity in this?

The Zionist movement was initially secular, though religious elements have been gaining a considerably greater role, particularly after the 1967 war and the onset of the occupation, which had a major impact on Israeli society and culture. That's particularly true in the military, a matter that has deeply concerned analysts of military affairs since the 1980s (Yoram Peri's warnings at the time were perceptive) and increasingly today. The Palestinian movements were also largely secular, though religious extremism is also growing—throughout the Muslim world, in fact, as secular initiatives are beaten back and the victims seek something else to grasp. Still, it would be quite misleading, I think, to regard it as a war of religion. Whatever one thinks of it, Zionism has been a settler-colonial movement, with all that that entails.

What do you think of the French law on secularity and conspicuous religious symbols? A step forward or backward on progress and universalism?

I don't think there should be laws forcing women to remove veils or preferred clothes when swimming. Secular values should, I think, be honored; among them, respect for individual choice, as long as it does not harm others. Secular values that should be respected are undermined when state power intrudes in areas that should be matters of personal choice. If Hasidic Jews choose to dress in black cloaks, white shirts, and black hats, with hair in orthodox style and with religious garb, that's not the state's business. Same when a Muslim woman decides to wear a scarf or go swimming in a "burkini."

Constructing Visions of "Perpetual Peace"

C. J. POLYCHRONIOU: Noam, the decline of democracy as a reflection of political apathy is evident in both the United States and in Europe, and the explanation provided in *Who Rules the World?* is that this phenomenon is linked to the fact that most people throughout Western societies are "convinced that a few big interests control policy."[1] This is obviously true, but wasn't this always the case? I mean, people always knew that policy making was in the hands of the elite, but this did not stop them in the past from seeking to influence political outcomes through the ballot box and other means. So, what specific factors might explain political apathy in our own age?

NOAM CHOMSKY: "Resignation" may be a better term than "apathy," and even that goes too far, I think.

Since the early 1980s, polls in the United States have shown that most people believe that the government is run by a few big interests looking out for themselves. I do not know of earlier polls, or polls in other countries, but it would not be surprising if the results are similar. The important question is: Are people motivated to do something about it? That depends on many factors, crucially including the means that they perceive to be available. It's the task of serious activists to help develop those means and encourage people to understand that they are available. Two hundred and fifty years

---

Originally published in *Truthout*, June 19, 2016

ago, in one of the first modern works of political theory, David Hume observed that "power is in the hands of the governed," if they only choose to exercise it, and ultimately, it is "by opinion only"—that is, by doctrine and propaganda—that they are prevented from exercising power. That can be overcome, and often has been.

Thirty-five years ago, political scientist Walter Dean Burnham identified "the total absence of a socialist or laborite mass party as an organized competitor in the electoral market" as a primary cause of the high rate of abstention in US elections. Traditionally, the labor movement and labor-based parties have played a leading role in offering ways to "influence political outcomes" within the electoral system and on the streets and shop floor. That capacity has declined significantly under neoliberal assault, which enhanced the bitter war waged against unions by the business classes throughout the postwar period.

In 1978, before Reagan's escalation of the attack against labor, United Auto Workers president Doug Fraser recognized what was happening—far too late—and criticized the "leaders of the business community" for having "chosen to wage a one-sided class war in this country—a war against working people, the unemployed, the poor, minorities, the very young and the very old, and even many in the middle class of our society," and for having "broken and discarded the fragile, unwritten compact previously existing during a period of growth and progress." The union leadership had placed their faith—partly for their own benefit as a labor bureaucracy—in a compact with owners and managers during the postwar growth and high profits period that had come to an end by the 1970s. By then, the powerful attack on labor had already taken a severe toll and it has gotten much more extreme since, particularly since the radically antilabor Reagan administration.

The Democrats, meanwhile, pretty much abandoned the working class. Independent political parties have been very marginal, and political activism, while widespread, has often sidelined class issues and offered little to the white working class, which is now drifting into the hands of their class enemy. In Europe, functioning democracy has steadily declined as major policy decisions are transferred to the Brussels bureaucracy of the EU, operating under the shadow of Northern banks. But there are many popular reactions, some self-destructive (racing into the hands of the class enemy) and others quite promising and productive, as we see in current political campaigns in the United States and Europe.

In your book, you refer to the "invisible hands of power." What is the exact meaning of this, and to what situations and circumstances can it be applied in order to understand domestic and global political developments?

I was using the phrase to refer to the guiding doctrines of policy formation, sometimes spelled out in the documentary record, sometimes easily detectable in ongoing events. There are many examples in international and domestic affairs. Sometimes the clouds are lifted by high-level disclosures or by significant historical events. The real nature of the Cold War, for example, was considerably illuminated when the Soviet Union collapsed and it was no longer possible to proclaim simply that the Russians are coming. That provided an interesting test of the real motives of policy formation, hidden by Cold War pretexts that were suddenly gone.

We learn from Bush I administration documents, for example, that we must keep intervention forces aimed at the Middle East, where the serious threats to our interests "could not be laid at the Kremlin's door," contrary to long deceit. Rather, the serious problems trace to "radical nationalism," the term regularly used for independent nationalism that is under control. That is actually a major theme of the Cold War, masked by posturing about the Great Enemy.

The fate of NATO is also revealing. It was constructed and maintained in alleged defense against the Russian hordes. By 1991, there were no more Russian hordes, no Warsaw Pact, and Mikhail Gorbachev was proposing a broad security system with no military pacts. What happened to NATO? It expanded to the East in violation of commitments to Gorbachev by President Bush I and secretary of state James Baker that appear to have been consciously intended to deceive him and to gain his acquiescence to a unified Germany within NATO, so recent archival work persuasively indicates.

To move to another domain, the free-market capitalism extolled in doctrine was illustrated by an IMF study of major banks, which showed that their profits derived mostly from an implicit taxpayer insurance policy.

Examples abound, and are highly instructive.

Since the end of World War II, capitalism throughout the West—and in fact throughout the globe—has managed to maintain and expand its domination not merely through political and psychological means but also through the use of the repressive apparatus of the state, including the military. Can you talk a little bit about this in connection with the theme of "who rules the world"?

The "mailed fist" (the threat of armed or overbearing force) is not lacking even within the most free societies. In the postwar United States, the most striking example is COINTELPRO, a program run by the national political police (FBI) to stamp out dissidence and activism over a broad range, reaching as far as political assassination (Black Panther organizer Fred Hampton). Massive incarceration of populations deemed superfluous for profit-making (largely African American, for obvious historical reasons) is yet another means.

Abroad, the fist is constantly wielded, directly or through clients. The Indochina wars are the most extreme case, the worst postwar twentieth-century crime, criticized in the mainstream as a "blunder," like the invasion of Iraq, the worst crime of the new century. One highly significant postwar example is the plague of violent repression that spread through Latin America after John F. Kennedy effectively shifted the mission of the Latin American military from "hemispheric defense" to "internal security," a euphemism for war against the population. There were horrendous effects throughout the hemisphere, reaching Central America with Reagan's murderous wars, mostly relying on the terrorist forces of client states.

While still the world's predominant power, there is no doubt that the United States is in decline. What are the causes and consequences of American decline?

US power peaked, at a historically unprecedented level, at the end of World War II. That couldn't possibly be sustained. It began to erode very soon with what is called, interestingly, "the loss of China" (the transformation of China into a communist nation in 1949). And the process continued with the reconstruction of industrial societies from wartime devastation and decolonization. One reflection of the decline is the shift of attitudes toward the UN. It was greatly admired when it was hardly more than an instrument of US power in the early postwar years, but increasingly came under attack as "anti-American" as it fell out of control—so far out of control that the United States has held the record in vetoes after 1970, when it joined Britain in support of the racist regime of Southern Rhodesia. By then, the global economy was tripartite: German-based Europe, Japan-based East Asia, and US-based North America.

In the military dimension, the United States has remained supreme. There are many consequences. One is resort to "coalitions of the willing" when international opinion overwhelmingly opposes US resort to violence, even among allies, as in the case of the invasion of Iraq. Another is "soft coups," as right now in Brazil, rather than support for neo-Nazi national security states, as was true in the not-distant past.

If the United States is still the world's first superpower, what country or entity do you consider to be the second superpower?

There is much talk of China as the emerging superpower. According to many analysts, it is poised to overtake the United States. There is no doubt of China's emerging significance in the world scene, already surpassing the United States economically by some measures (though far below per capita). Militarily, China is far weaker; confrontations are taking place in coastal waters near China, not in the Caribbean or off the coast of California. But China faces very serious internal problems—labor repression and protest, severe ecological threats, demographic decline in work force, and others. And the economy, while booming, is still highly dependent on the more advanced industrial economies at its periphery and the West, though that is changing. In some high-tech domains, such as design and development of solar panels, China seems to have the world lead. As China is hemmed in from the sea, it is compensating by extending westward, reconstructing something like the old silk roads in a Eurasian system largely under Chinese influence and soon to reach Europe.

You have been arguing for a long time now that nuclear weapons pose one of the two greatest threats to humankind. Why are the major powers

so reluctant to abolish nuclear weapons? Doesn't the very existence of these weapons pose a threat to the existence of the "masters of the universe" themselves?

It is quite remarkable to see how little concern top planners show for the prospects of their own destruction—not a novelty in world affairs (those who initiated wars often ended up devastated) but now on a hugely different scale. We see that from the earliest days of the atomic age. The United States at first was virtually invulnerable, though there was one serious threat on the horizon: ICBMs (intercontinental ballistic missiles) with hydrogen bomb warheads. Archival research has now confirmed what was surmised earlier: there was no plan, not even a thought, of reaching a treaty agreement that would have banned these weapons, though there is good reason to believe that it might have been feasible. The same attitudes prevail right to the present, where the vast buildup of forces right at the traditional invasion route into Russia is posing a serious threat of nuclear war.

Planners explain quite lucidly why it is so important to keep these weapons. One of the clearest explanations is in a partially declassified Clinton-era document issued by the Strategic Command (STRATCOM), which is in charge of nuclear weapons policy and use. The document is called *Essentials of Post–Cold War Deterrence*; the term "deterrence," like "defense," is a familiar Orwellism referring to coercion and attack. The document explains that "nuclear weapons always cast a shadow over any crisis or conflict," and must therefore be available, at the ready. If the adversary knows we have them, and might use them, they may back down—a regular feature of Kissingerian diplomacy. In that sense, nuclear weapons are constantly being used, a point that Dan Ellsberg has insistently made, just as we are using a gun when we rob a store but don't actually shoot. One section of the report is headed "Maintaining Ambiguity." It advises that "planners should not be too rational about determining . . . what the opponent values the most," which must be targeted.

"That the US may become irrational and vindictive if its vital interests are attacked should be a part of the national persona we project," the report says, adding that it is "beneficial" for our strategic posture if "some elements may appear to be potentially 'out of control.'" Nixon's madman theory, except this time clearly articulated in an internal planning document, not merely a recollection by an adviser (Haldeman, in the Nixon case).

Like other early post–Cold War documents, this one has been virtually ignored. (I've referred to it a number of times, eliciting no notice that I'm aware of.) The neglect is quite interesting. Simple logic suffices to show that the documentary record after the alleged Russian threat disappeared would be highly illuminating as to what was actually going on before.

The Obama administration has made some openings toward Cuba. Do you anticipate an end to the embargo any time soon?

The embargo has long been opposed by the entire world, as the annual votes on the embargo at the UN General Assembly reveal. By now the United States is supported only by Israel. Before, it could sometimes count on a Pacific island or some other dependency. Of course Latin America is completely opposed. More interestingly, major sectors of US capital have long been in favor of normalization of relations, as public opinion has been: agribusiness, pharmaceuticals, energy, tourism, and others. It is normal for public opinion to be ignored, but dismissing powerful concentrations of the business world tells us that really significant "reasons of state" are involved. We have a good sense from the internal record about what these interests are.

From the Kennedy years until today there has been outrage over Cuba's "successful defiance" of US policies going back to the Monroe Doctrine, which signaled the intention to control the hemisphere. The goal was not realizable because of relative weakness, just as the British deterrent prevented the United States from attaining its first "foreign policy" objective, the conquest of Cuba, in the 1820s (here the term "foreign policy" is used in the conventional sense, which adheres to what historian of imperialism Bernard Porter calls "the salt water fallacy": conquest becomes imperial only when it crosses salt water, so the destruction of the Indian nations and the conquest of half of Mexico were not "imperialism"). The United States did achieve its objective in 1898, intervening to prevent Cuba's liberation from Spain and converting it into a virtual colony.

Washington has never reconciled itself to Cuba's intolerable arrogance of achieving independence in 1959—partial, since the United States refused to return the valuable Guantanamo Bay region, taken by "treaty" at gunpoint in 1903 and not returned despite the requests of the government of Cuba. In passing, it might be recalled that by far the worst human rights violations in

Cuba take place in this stolen territory, to which the United States has a much weaker claim than Russia does to Crimea, also taken by force.

But to return to the question, it is hard to predict whether the United States will agree to end the embargo short of some kind of Cuban capitulation to US demands going back almost two hundred years.

How do you assess and evaluate the historical significance and impact of the Cuban revolution in world affairs and toward the realization of socialism?

The impact on world affairs was extraordinary. For one thing, Cuba played a very significant role in the liberation of West and South Africa. Its troops beat back a US-supported South African invasion of Angola and compelled South Africa to abandon its attempt to establish a regional support system and to give up its illegal hold on Namibia. The fact that Black Cuban troops defeated the South Africans had an enormous psychological impact both in white and Black Africa. A remarkable exercise of dedicated internationalism, undertaken at great risk from the reigning superpower, which was the last supporter of apartheid South Africa, and entirely selfless. Small wonder that when Nelson Mandela was released from prison, one of his first acts was to declare:

> During all my years in prison, Cuba was an inspiration and Fidel Castro a tower of strength. . . . [Cuban victories] destroyed the myth of the invincibility of the white oppressor [and] inspired the fighting masses of South Africa . . . a turning point for the liberation of our continent—and of my people—from the scourge of apartheid . . . What other country can point to a record of greater selflessness than Cuba has displayed in its relations to Africa?

Cuban medical assistance in poor and suffering areas is also quite unique.

Domestically, there were very significant achievements, among them simply survival in the face of US efforts to bring "the terrors of the earth" to Cuba (historian Arthur Schlesinger's phrase, in his biography of Robert Kennedy, who was assigned this task as his highest priority) and the fierce embargo. Literacy campaigns were highly successful, and the health system is justly renowned. There are serious human rights violations and restrictions of political and personal freedoms. How much is attributable to the external attack and how much to independent policy choices, one can debate—but for Americans to condemn violations without full recognition of their own massive responsibility gives hypocrisy a new meaning.

Does the United States remain the world's leading supporter of terrorism?

A review of several recent books on Obama's global assassination (drone) campaign in the *American Journal of International Law* concludes that there is a "persuasive case" that the campaign is "unlawful": "U.S. drone attacks generally violate international law, worsen the problem of terrorism, and transgress fundamental moral principles"—a judicious assessment, I believe. The details of the cold and calculated presidential killing machine are harrowing, as is the attempt at legal justification, such as the stand of Obama's Justice Department on "presumption of innocence," a foundation stone of modern law tracing back to the Magna Carta eight hundred years ago. As the stand was explained in the *New York Times*, "Mr. Obama embraced a disputed method for counting civilian casualties that did little to box him in. It, in effect, counts all military-age males in a strike zone as combatants, according to several administration officials, unless there is explicit intelligence posthumously proving them innocent"—post-assassination. In large areas of tribal Pakistan and Yemen, and elsewhere, populations are traumatized by the fear of sudden murder from the skies at any moment. The distinguished anthropologist Akbar Ahmed, with long professional and personal experience with the tribal societies that are under attack all over the world, forcefully recounts how these murderous assaults elicit dedication to revenge—not very surprisingly. How would we react?

These campaigns alone, I think, secure the trophy for the United States.

Historically, under capitalism, plundering the poor and the natural resources of weak nations has been the favorite hobby of both the rich and of imperial states. In the past, the plundering was done mostly through outright physical exploitation means and military conquest. How have the means of exploitation changed under financial capitalism?

Secretary of state John Foster Dulles once complained to President Eisenhower that the Communists have an unfair advantage. They can "appeal directly to the masses" and "get control of mass movements, something we have no capacity to duplicate. The poor people are the ones they appeal to and they have always wanted to plunder the rich." It's not easy to sell the principle that the rich have a right to plunder the poor.

It's true that the means have changed. The international "free trade agreements" (FTAs) are a good example, including those now being negotiated—mostly in secret from populations, but not from the corporate lawyers and lobbyists who are writing the details. The FTAs reject "free trade": they are highly protectionist, with onerous patent regulations to guarantee exorbitant profits for the pharmaceutical industry, media conglomerates, and others, as well as protection for affluent professionals, unlike working people, who are placed in competition with all of the world, with obvious consequences. The FTAs are to a large extent not even about trade; rather, about investor rights, such as the rights of corporations (not, of course, mere people of flesh and blood) to sue governments for actions that might reduce potential profits of foreign investors, like environmental or health and safety regulations. Much of what is called "trade" doesn't merit that term, for example, production of parts in Indiana, assembly in Mexico, sale in California, all basically within a command economy, a megacorporation. Flow of capital is free. Flow of labor is anything but, violating what Adam Smith recognized to be a basic principle of free trade: free circulation of labor. And to top it off, the FTAs are not even agreements, at least if people are considered to be members of democratic societies.

Is this to say that we now live in a postimperialist age?

Seems to me just a question of terminology. Domination and coercion take many and varied forms, as the world changes.

We have seen in recent years several so-called progressive leaders march to power through the ballot box only to betray their vows to the people the moment they took office. What means or mechanisms should be introduced in truly democratic systems to ensure that elected officials do not betray the trust of the voters? For example, the ancient Athenians had conceived of something called "the right to recall," which in the nineteenth century became a critical although little-known element in the political project for future social and political order of certain

socialist movements. Are you in favor of reviving this mechanism as a critical component of real, sustainable democracy?

I think a strong case can be made for right of recall in some form, buttressed by capacities for free and independent inquiry to monitor what elected representatives are doing. The great achievement of Chelsea Manning, Julian Assange, Edward Snowden, and other contemporary "whistleblowers" is to serve and advance these fundamental rights of citizens. The reaction by state authorities is instructive. As is well known, the Obama administration has broken all records in punishment of whistleblowers. It is also remarkable to see how intimidated Europe is. We saw that dramatically when Bolivian president Evo Morales's plane flew home from a visit to Moscow, and European countries were in such terror of Washington that they would not let the plane cross their airspace, in case it might be carrying Edward Snowden, and when the plane landed in Austria it was searched by police in violation of diplomatic protocol.

Could an act of terrorism against leaders who blatantly betrayed the trust of voters ever be justified?

"Ever" is a strong word. It is hard to conjure up realistic circumstances. The burden of proof for any resort to violence should be very heavy, and this case would seem extremely hard to justify.

With human nature being what it is, and individuals clearly having different skills, abilities, drives, and aspirations, is a truly egalitarian society feasible and/or desirable?

Human nature encompasses saints and sinners, and each of us has all of these capacities. I see no conflict at all between an egalitarian vision and human variety. One could, perhaps, argue that those with greater skills and talents are already rewarded by the ability to exercise them, so they merit less external reward—though I don't argue this. As for the feasibility of more just and free social institutions and practices, we can never be certain in advance, and can only keep trying to press the limits as much as possible, with no clear reason that I can see to anticipate failure.

In your view, what would constitute a decent society and what form of a world order would be needed to eliminate completely questions about who rules the world?

We can construct visions of "perpetual peace," carrying forward the Kantian project, and of a society of free and creative individuals not subjected to hierarchy, domination, arbitrary rule and decision. In my own view—respected friends and comrades in struggle disagree—we do not know enough to spell out details with much confidence, and can anticipate that considerable experimentation will be necessary along the way. There are very urgent immediate tasks, not least dealing with literal questions of survival of organized human societies, questions that have never risen before in human history but are inescapable right now. And there are many other tasks that demand immediate and dedicated work. It makes good sense to keep in mind longer-term aspirations as guidelines for immediate choices, recognizing as well that the guidelines are not immutable. That leaves us plenty to do.

It Is All Working Quite Well for the Rich, Powerful*

C. J. POLYCHRONIOU: Neoliberal ideology claims that the government is a problem, society does not exist, and individuals are responsible for their own fate. Yet, big business and the rich rely, as ever, on state intervention to maintain their hold over the economy and to enjoy a bigger slice of the economic pie. Is neoliberalism a myth, merely an ideological construct?

NOAM CHOMSKY: The term "neoliberal" is a bit misleading. The doctrines are neither new nor liberal. As you say, big business and the rich rely extensively on what economist Dean Baker calls "the conservative nanny state" that they nourish. That is dramatically true of financial institutions. A recent IMF study attributes the profits of the big banks almost entirely to the implicit government insurance policy ("too big to fail"), not just the widely publicized bailouts but access to cheap credit, favorable ratings because of the state guarantee, and much else. The same is true of the productive economy. The IT revolution, now its driving force, relied very heavily on state-based R&D, procurement, and other devices. That pattern goes back to early English industrialization.

However, neither "neoliberalism," nor its earlier versions as "liberalism," have been myths, certainly not for their victims. Economic historian Paul Bairoch is only one of many who have shown that "the Third World's compulsory economic liberalism in the nineteenth century is a major

*Coauthored with Anastasia Giamali; originally published in *Truthout*, December 8, 2013

element in explaining the delay in its industrialization," in fact, its "deindustrialization," a story that continues to the present under various guises.

In brief, the doctrines are, to a substantial extent, a "myth" for the rich and powerful, who craft many ways to protect themselves from market forces, but not for the poor and weak, who are subjected to their ravages.

What explains the supremacy of market-centric rule and predatory finance in an era that has experienced the most destructive crisis of capitalism since the Great Depression?

The basic explanation is the usual one: it is all working quite well for the rich and powerful. In the United States, for example, tens of millions are unemployed, unknown millions have dropped out of the workforce in despair, and incomes as well as conditions of life have largely stagnated or declined. But the big banks, which were responsible for the latest crisis, are bigger and richer than ever. Corporate profits are breaking records, wealth beyond the dreams of avarice is accumulating among those who count, and labor is severely weakened by union busting and "growing worker insecurity," to borrow the term Alan Greenspan used in explaining the grand success of the economy he managed, when he was still "St. Alan"—perhaps the greatest economist since Adam Smith, before the collapse of the structure he had administered, along with its intellectual foundations. So what is there to complain about?

The growth of financial capital is related to the decline in the rate of profit in industry and the new opportunities to distribute production more widely to places where labor is more readily exploited and constraints on capital are weakest—while profits are distributed to places with lowest tax rates ("globalization"). The process has been abetted by technological developments that facilitate the growth of an "out-of-control financial sector," which "is eating out the modern market economy [that is, the productive economy] from inside, just as the larva of the spider wasp eats out the host in which it has been laid," to borrow the evocative phrase of Martin Wolf of the *Financial Times*, probably the most respected financial correspondent in the English-speaking world.

That aside, as noted, the "market-centric rule" imposes harsh discipline on the many, but the few who count protect themselves from it effectively.

What do you make of the argument about the dominance of a transnational elite and the end of the nation-state, especially since its proponents claim that this New World Order is already upon us?

There's something to it, but it shouldn't be exaggerated. Multinationals continue to rely on the home state for protection, economic and military, and substantially for innovation as well. The international institutions remain largely under the control of the most powerful states, and in general the state-centric global order remains reasonably stable.

Europe is moving ever closer to the end of the "social contract." Is this a surprising development for you?

In an interview, Mario Draghi informed the *Wall Street Journal* that "the Continent's traditional social contract"—perhaps its major contribution to contemporary civilization—"is obsolete" and must be dismantled. And he is one of the international bureaucrats who is doing most to protect its remnants. Business has always disliked the social contract. Recall the euphoria in the business press when the fall of "communism" offered a new work force—educated, trained, healthy, and even blond and blue-eyed—that could be used to undercut the "luxurious lifestyle" of Western workers. It is not the result of inexorable forces, economic or other, but a policy design based on the interests of the designers, who are rather more likely to be bankers and CEOs than the janitors who clean their offices.

One of the biggest problems facing many parts of the advanced capitalist world today is the debt burden, public and private. In the peripheral nations of the Eurozone, in particular, debt is having catastrophic social effects as the "people always pay," as you have pointedly argued in the past. For the benefit of today's activists, would you explain in what sense debt is "a social and ideological construct?"

There are many reasons. One was captured well by a phrase of the US executive director of the IMF, Karen Lissakers, who described the institution as "the credit community's enforcer." In a capitalist economy, if you lend me money and I can't pay you back, it's your problem: you cannot demand that my neighbors pay the debt. But since the rich and powerful protect themselves from market discipline, matters work differently when a big bank lends money to risky borrowers, hence at high interest and profit, and at some point they cannot pay. Then the "the credit community's enforcer" rides to the rescue, ensuring that the debt is paid, with liability transferred to the general public by structural adjustment programs, austerity, and the like. When the rich don't like to pay such debts, they can declare them to be "odious," hence invalid: imposed on the weak by unfair means. A huge amount of debt is "odious" in this sense, but few can appeal to powerful institutions to rescue them from the rigors of capitalism.

There are plenty of other devices. J. P. Morgan Chase has just been fined $13 billion (half of it tax-deductible) for what should be regarded as criminal behavior in fraudulent mortgage schemes, from which the usual victims suffer under hopeless burdens of debt.

The inspector-general of the US government bailout program, Neil Barofsky, pointed out that it was officially a legislative bargain: the banks that were the culprits were to be bailed out, and their victims, people losing their homes, were to be given some limited protection and support. As he explains, only the first part of the bargain was seriously honored, and the plan became a "giveaway to Wall Street executives"—to the surprise of no one who understands "really existing capitalism."

The list goes on.

In the course of the crisis, Greeks have been portrayed around the globe as lazy and corrupt tax evaders who merely like to demonstrate. This view has become mainstream. What are the mechanisms used to persuade public opinion? Can they be tackled?

The portrayals are presented by those with the wealth and power to frame the prevailing discourse. The distortion and deceit can be confronted only by undermining their power and creating organs of popular power, as in all other cases of oppression and domination.

What is your view about what is happening in Greece, particularly with regard to the constant demands by the "troika" and Germany's unyielding desire to advance the cause of austerity?

It appears that the ultimate aim of the German demands from Athens, under the management of the debt crisis, is the capture of whatever is of value in Greece. Some people in Germany appear to be intent on imposing conditions of virtual economic slavery on the Greeks.

It is rather likely that the next government in Greece will be a government of the Coalition of the Radical Left. What should be its approach toward the European Union and Greece's creditors? Also, should a left government be reassuring toward the most productive sectors of the capitalist class, or should it adopt the core components of a traditional workerist-populist ideology?

These are hard practical questions. It would be easy for me to sketch what I would like to happen, but given existing realities, any course followed has risks and costs. Even if I were in a position to assess them properly—I am not—it would be irresponsible to urge policy without serious analysis and evidence.

Capitalism's appetite for destruction was never in doubt, but in your recent writings you pay increasing attention to environmental destruction. Do you really think human civilization is at stake?

I think decent human survival is at stake. The earliest victims are, as usual, the weakest and most vulnerable. That much has been evident even in the global summit on climate change that just concluded in Warsaw, with little outcome. And there is every reason to expect that to continue. A future historian—if there is one—will observe the current spectacle with amazement. In the lead in trying to avert likely catastrophe are the so-called primitive societies: First Nations in Canada, Indigenous people in South America, and so on, throughout the world. We see the struggle for environmental salvage and protection taking place today in Greece, where the residents of Skouries in Chalkidiki are putting up a heroic resistance both against the predatory aims of Eldorado Gold and the police forces that have been mobilized by the Greek state in support of the multinational company.

Those enthusiastically leading the race to fall off the cliff are the richest and most powerful societies, with incomparable advantages, like the United

States and Canada. Just the opposite of what rationality would predict—apart from the lunatic rationality of "really existing capitalist democracy."

The United States remains a world empire and, by your account, operates under the "Mafia principle," meaning that the godfather does not tolerate "successful defiance." Is the American empire in decline, and, if so, does it pose yet a greater threat to global peace and security?

US global hegemony reached a historically unparalleled peak in 1945, and has been declining steadily since. Though it still remains very great and though power is becoming more diversified, there is no single competitor in sight. The traditional Mafia principle is constantly invoked, but ability to implement it is more constrained. The threat to peace and security is very real. To take just one example, President Obama's drone campaign is by far the most vast and destructive terrorist operation now under way. The United States and its Israeli client violate international law with complete impunity, for example, by threats to attack Iran ("all options are open") in violation of core principles of the UN Charter. The most recent US Nuclear Posture Review (2010) is more aggressive in tone than its predecessors, a warning not to be ignored. Concentration of power rather generally poses dangers, in this domain as well.

Regarding the Israeli–Palestinian conflict, you have said all along that the one-state/two-state debate is irrelevant.

The one-state/two-state debate is irrelevant because one state is not an option. It is worse than irrelevant: it is a distraction from the reality.

The actual options are either (1) two states or (2) a continuation of what Israel is now doing with US support: keeping Gaza under a crushing siege, separated from the West Bank; systematically taking over what it finds of value in the West Bank while integrating it more closely to Israel, taking over areas with not many Palestinians; and quietly expelling those who are there. The contours are quite clear from the development and expulsion programs.

Given option (2), there's no reason why Israel or the United States should agree to the one-state proposal, which also has no international support anywhere else. Unless the reality of the evolving situation is recognized, talk about one state (civil rights/antiapartheid struggle, "demographic problem," and so on) is just a diversion, implicitly lending support to option (2). That's the essential logic of the situation, like it or not.

You have said that elite intellectuals are the ones that mainly tick you off. Is this because you fuse politics with morality?

Elite intellectuals, by definition, have a good deal of privilege. Privilege provides options and confers responsibility. Those more privileged are in a better position to obtain information and to act in ways that will affect policy decisions. Assessment of their role follows at once.

It's true that I think that people should live up to their elementary moral responsibilities, a position that should need no defense. And the responsibilities of someone in a more free and open society are, again obviously, greater than those who may pay some cost for honesty and integrity. If commissars in Soviet Russia agreed to subordinate themselves to state power, they could at least plead fear in extenuation. Their counterparts in more free and open societies can plead only cowardice.

Michel Gondry's animated documentary *Is the Man Who Is Tall Happy?* has just been released in selected theaters in New York City and other major cities in the United States after having received rave reviews. Did you see the movie? Were you pleased with it? [*Ed. Note*: Is the Man Who Is Tall Happy? *is based on a series of interviews featuring Noam Chomsky.*]

I saw it. Gondry is really a great artist. The movie is delicately and cleverly done and manages to capture some important ideas (often not understood even in the field) in a very simple and clear way, also with personal touches that seemed to me very sensitive and thoughtful.

Can Civilization Survive "Really Existing Capitalism"?

C. J. POLYCHRONIOU: In a nationally televised address on the eve of the thirteenth anniversary of the September 11, 2001, attacks on the United States, Obama announced to the American people and the rest of the world that the United States is going back to war in Iraq, this time against the self-proclaimed Islamic State of Iraq and Syria (ISIS). Is Iraq an unfinished business of the US invasion of 2003, or is the situation there merely the inevitable outcome of the strategic agenda of the Empire of Chaos?

NOAM CHOMSKY: "Inevitable" is a strong word, but the appearance of ISIS and the general spread of radical jihadism is a fairly natural outgrowth of Washington wielding its sledgehammer at the fragile society of Iraq, which was barely hanging together after a decade of US-UK sanctions so onerous that the respected international diplomats who administered them via the UN both resigned in protest, charging that they were "genocidal."

One of the most respected mainstream US Middle East analysts, former CIA operative Graham Fuller, recently wrote: "I think the United States is one of the key creators of [ISIS]. The United States did not plan the formation of ISIS, but its destructive interventions in the Middle East and the war in Iraq were the basic causes of the birth of ISIS."

He is correct, I think. The situation is a disaster for the United States but

Originally published in *Truthout*, October 1, 2014

is a natural result of its invasion. One of the grim consequences of US-UK aggression was to inflame sectarian conflicts that are now tearing Iraq to shreds, and have spread over the whole region, with awful consequences.

ISIS seems to represent a new jihadist movement, with greater inherent tendencies toward barbarity in the pursuit of its mission to reestablish an Islamic caliphate, yet apparently more able to recruit young radical Muslims from the heart of Europe, and even as far as Australia, than al-Qaeda itself. In your view, why has religious fanaticism become the driving force behind so many Muslim movements around the world?

Like Britain before it, the United States has tended to support radical Islam and to oppose secular nationalism, which both imperial states have regarded as more threatening to their goals of domination and control. When secular options are crushed, religious extremism often fills the vacuum. Furthermore, the primary US ally over the years, Saudi Arabia, is the most radical Islamist state in the world and also a missionary state, which uses its vast oil resources to promulgate its extremist Wahhabi/Salafi doctrines by establishing schools, mosques, and in other ways, and has also been the primary source for the funding of radical Islamist groups, along with Gulf Emirates—all US allies.

It's worth noting that religious fanaticism is spreading in the West as well, as democracy erodes. The United States is a striking example. There are not many countries in the world where the large majority of the population believes that God's hand guides evolution, and almost half of these think that the world was created a few thousand years ago. And as the Republican Party has become so extreme in serving wealth and corporate power that it cannot appeal to the public on its actual policies, it has been compelled to rely on these sectors as a voting base, giving them substantial influence on policy.

The United States committed major war crimes in Iraq, but the acts of violence committed these days against civilians in the country, particularly against children and people from various ethnic and religious communities, is also simply appalling. Given that Iraq exhibited its longest

stretch of political stability under Saddam Hussein, what didactic lessons should one draw from today's extremely messy situation in that part of the world?

The most elementary lesson is that it is wise to adhere to civilized norms and international law. The criminal violence of rogue states like the United States and UK is not guaranteed to have catastrophic consequences, but we can hardly claim to be surprised when it does.

US attacks against ISIS's bases in Syria without the approval and collaboration of the Syrian regime of Bashar al-Assad would constitute a violation of international law, claimed Damascus, Moscow, and Tehran before the start of bombing. However, isn't it the case that the destruction of ISIS's forces in Syria would further strengthen the Syrian regime? Or is it that the Assad regime is afraid it will be next in line?

The Assad regime has been rather quiet. It has not, for example, appealed to the Security Council to act to terminate the attack, which is, undoubtedly, in violation of the UN Charter, the foundation of modern international law (and if anyone cares, part of the "supreme law of the land" in the United States, under the Constitution). Assad's murderous regime doubtless can see what the rest of the world does: the US attack on ISIS weakens its main enemy.

In addition to some Western nations, Arab states have also offered military support to US attacks against ISIS in Iraq and Syria. Is this a case of one form of Islamic fundamentalism (Saudi Arabia, for example) exhibiting fear of another form of Islamic fundamentalism (ISIS)?

As the *New York Times* accurately reported, the support is "tepid." The regimes surely fear ISIS, but it apparently continues to draw financial support from wealthy donors in Saudi Arabia and the Emirates, and its ideological roots, as I mentioned, are in Saudi radical Islamic extremism, which has not abated.

Life in Gaza has returned to normalcy after Hamas and Israel agreed to a cease-fire. For how long?

I would hesitate to use the term "normalcy." The latest onslaught was even more vicious than its predecessors, and its impact is horrendous. The Egyptian military dictatorship, which is bitterly anti-Hamas, is also adding to the tragedy.

What will happen next? There has been a regular pattern since the first such agreement was reached between Israel and the Palestinian Authority in November 2005. It called for "a crossing between Gaza and Egypt at Rafah for the export of goods and the transit of people, continuous operation of crossings between Israel and Gaza for the import/export of goods, and the transit of people, reduction of obstacles to movement within the West Bank, bus and truck convoys between the West Bank and Gaza, the building of a seaport in Gaza, [and the] re-opening of the airport in Gaza" that Israeli bombing had demolished.

Later agreements have been variants on the same themes, the current one as well. Each time, Israel has disregarded the agreements while Hamas has lived up to them (as Israel concedes) until some Israeli escalation elicits a Hamas response, which gives Israel another opportunity to "mow the lawn," in its elegant phrase. The interim periods of "quiet" (meaning one-way quiet) allow Israel to carry forward its policies of taking over whatever it values in the West Bank, leaving Palestinians in dismembered cantons. All, of course, with crucial US support: military, economic, diplomatic, and ideological, in framing the issues in accord with Israel's basic perspective.

That, indeed, was the purpose of Israel's "disengagement" from Gaza in 2005—while remaining the occupying power, as recognized by the world (apart from Israel), even the United States. The purpose was outlined candidly by the architect and chief negotiator of the "disengagement," Prime Minister Sharon's close associate, Dov Weissglass. He informed the press:

> The significance of the disengagement plan is the freezing of the peace process. And when you freeze that process, you prevent the establishment of a Palestinian state, and you prevent a discussion on the refugees, the borders and Jerusalem. Effectively, this whole package called the Palestinian state, with all that it entails, has been removed indefinitely from our agenda. And all this with authority and permission. All with a [US] presidential blessing and the ratification of both houses of Congress.

That pattern has been reiterated over and over, and it seems that it is being reenacted today. However, some knowledgeable Israeli commentators have suggested that Israel might finally relax its torture of Gaza. Its illegal takeover of much of the West Bank (including Greater Jerusalem) has proceeded so far that Israeli authorities might anticipate that it is irreversible. And they now have a cooperative ally in the brutal military dictatorship in Egypt. Furthermore, the rise of ISIS and the general shattering of the region have improved the tacit alliance with the Saudi dictatorship and possibly others. Conceivably, Israel might depart from its extreme rejectionism, though for now, the signs do not look auspicious.

The latest Israeli carnage in Gaza stirred public sentiment around the world increasingly against the state of Israel. To what extent is the unconditional support rendered by the United States toward Israel the outplay of domestic political factors, and under what conditions do you see a shift in Washington's policy toward Tel Aviv?

There are very powerful domestic factors. One illustration was given right in the midst of the latest Israeli assault. At one point, Israeli weapons seemed to be running low, and the United States kindly supplied Israel with more advanced weapons, which enabled it to carry the onslaught further. These weapons were taken from the stocks that the United States pre-positions in Israel, for eventual use by US forces, one of many indications of the very close military connections that go back many years. Intelligence interactions are even better established. Israel is also a favored location for US investors, not just in its advanced military economy. There is a huge voting bloc of evangelical Christians that is fanatically pro-Israel. There is also an effective Israel lobby, which is often pushing an open door—and which quickly backs down when it confronts US power, not surprisingly.

There are, however, shifts in popular sentiments, particularly among younger people, including the Jewish community. I experience that personally, as do others. Not long ago I literally had to have police protection when I spoke on these topics on college campuses, even my own university. That has greatly changed. By now Palestine solidarity is a major commitment on many campuses. Over time, these changes could combine with some other factors to lead to a change of US policy. It's happened before. But it will take hard, serious, dedicated work.

What are the aims and the objectives of US policy in Ukraine, other than stirring up trouble and then letting other forces do the dirty work?

Immediately after the fall of the Berlin Wall and the subsequent collapse of the USSR, the United States began seeking to extend its dominance, including NATO membership, over the regions released from Russian control—in violation of verbal promises to Gorbachev, whose protests were dismissed. Ukraine is surely the next ripe fruit that the United States hopes to pluck from the tree.

Doesn't Russia have a legitimate concern over Ukraine's potential alliance with NATO?

A very legitimate concern, over the expansion of NATO generally. This is so obvious that it is even the topic of the lead article in the current issue of the major establishment journal, *Foreign Affairs*, by international relations scholar John Mearsheimer. He observes that the United States is at the root of the current Ukraine crisis.

Looking at the current situation in Iraq, Syria, Libya, Nigeria, Ukraine, the China Sea, and even in parts of Europe, Zbigniew Brzezinski's recent comment on MSNBC, "We are facing a kind of dynamically spreading chaos in parts of the world," seems rather apropos. How much of this development is related to the decline of a global hegemon and to the balance of power that existed in the era of the Cold War?

US power reached its peak in 1945 and has been rather steadily declining ever since. There have been many changes in recent years. One is the rise of China as a major power. Another is Latin America's breaking free of imperial control (for the last century, US control) for the first time in five hundred years. Related to these developments is the rise of the BRICS bloc (Brazil, Russia, India, China, South Africa) and the Shanghai Cooperation Organization, based in China and including India, Pakistan, the Central Asian states, and others.

But the United States remains the dominant global power, by a large measure.

Last month marked the sixty-ninth anniversary of the US atomic bombing of the cities of Hiroshima and Nagasaki in Japan, yet nuclear disarmament remains a chimera. In a recent article of yours, you underscored the point that we are merely lucky to have avoided a nuclear war so far. Do you think, then, that it's a matter of time before nuclear weapons fall into the hands of terrorist groups?

Nuclear weapons are already in the hands of terrorist groups: state terrorists, the United States primary among them. It's conceivable that weapons of mass destruction might also fall into the hands of "retail terrorists," greatly enhancing the enormous dangers to survival.

Since the late 1970s, most advanced economies have returned to predatory capitalism. As a result, income and wealth inequality have reached spectacular heights, poverty is becoming entrenched, unemployment is skyrocketing and standards of living are declining. In addition, "really existing capitalism" is causing mass environmental damage and destruction, which, along with the population explosion, is leading us to an unmitigated global disaster. Can civilization survive really existing capitalism?

First, let me say that what I have in mind by the term "really existing capitalism" is what really exists and what is called "capitalism." The United States is the most important case, for obvious reasons. The term "capitalism" is vague enough to cover many possibilities. It is commonly used to refer to the US economic system, which receives substantial state intervention, ranging from creative innovation to the "too-big-to-fail" government insurance policy for banks, and which is highly monopolized, further limiting market reliance.

It's worth bearing in mind the scale of the departures of "really existing capitalism" from official "free-market capitalism." To mention only a few examples, in the past twenty years, the share of profits of the two hundred largest enterprises has risen sharply, carrying forward the oligopolistic character of the US economy. This directly undermines markets, avoiding price wars through efforts at often meaningless product differentiation through massive advertising, which is itself dedicated to undermining markets in the official sense, based on informed consumers making rational choices. Computers and the Internet, along with other basic components of the IT revolution, were largely in the state sector (R&D, subsidy, procurement, and other devices) for decades before they were handed over to private enterprise for adaptation to commercial markets and profit. The government insurance policy, which provides big banks with enormous advantages, has been roughly estimated by economists and the business press to be perhaps on the order of as much as $80 billion a year. However, a recent study by the IMF indicates—to quote the business press—that perhaps "the largest US banks aren't really profitable at all," adding that "the billions of dollars they allegedly earn for their shareholders were almost entirely a gift from US taxpayers."

In a way, all of this explains the economic devastation produced by contemporary capitalism that you underscore in your question above. Really existing capitalist democracy—RECD for short (pronounced "wrecked")—is radically incompatible with democracy. It seems to me unlikely that civilization can survive really existing capitalism and the sharply attenuated democracy that goes along with it. Could functioning democracy make a difference? Consideration of nonexistent systems can only be speculative, but I think there's some reason to think so. Really existing capitalism is a human creation, and can be changed or replaced.

Your book *Masters of Mankind*, which came out in September 2014 from Haymarket Books, is a collection of essays written between 1969 and 2013. The world has changed a great deal during this period, so my question is this: Has your understanding of the world changed over time, and, if so, what have been the most catalytic events in altering your perspective about politics?

My understanding of the world has changed over time as I've learned a lot more about the past, and ongoing events regularly add new critical materials. I can't really identify single events or people. It's cumulative, a constant process of rethinking in the light of new information and more consideration of what I hadn't properly understood. However, hierarchical and arbitrary power remains at the core of politics in our world and the source of all evils.

In a recent exchange we had, I expressed my pessimism about the future of our species. You replied by saying "I share your conviction, but keep remembering the line I've occasionally quoted from the Analects, defining the 'exemplary person'—presumably the master himself: 'the one who keeps trying, though he knows there is no hope.'" Is the situation as dire as that?

We cannot know for sure. What we do know, however, is that if we succumb to despair we will help ensure that the worst will happen. And if we grasp the hopes that exist and work to make the best use of them, there might be a better world.

Not much of a choice.

Part II

Vitality and the vote

United States, health metrics against swing to Donald Trump, by county

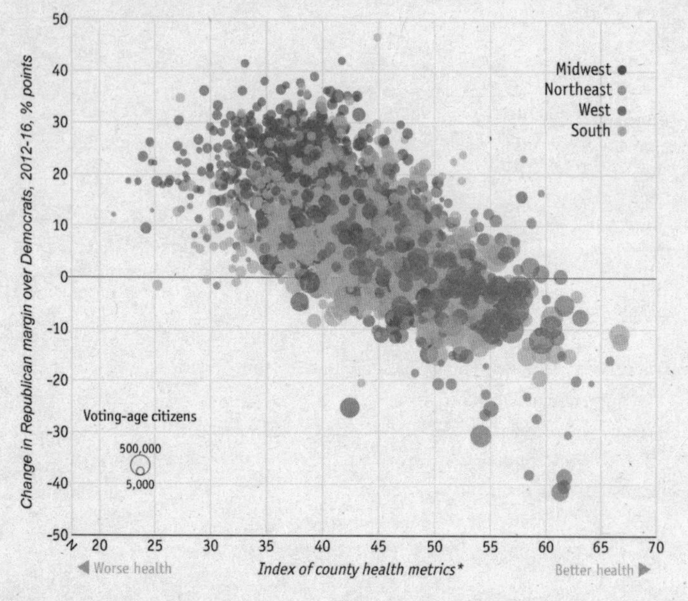

Sources: Atlas of US Presidential Elections; Census Bureau; IPUMS; Institute for Health Metrics and Evaluation; *The Economist*

*Weighted index of obesity, diabetes, heavy drinking, physical exercise and life expectancy, 2010-12

America in the Trump Era

C. J. POLYCHRONIOU: Noam, I want to start by asking you to reflect on the following: Trump won the presidential election even though he lost the popular vote. In this context, if "one person, one vote" is a fundamental principle behind every legitimate model of democracy, what type of a democracy prevails in the United States, and what will it take to undo the anachronism of the Electoral College?

NOAM CHOMSKY: The Electoral College was originally supposed to be a deliberative body drawn from educated and privileged elites. It would not necessarily respond to public opinion, which was not highly regarded by the founders, to put it mildly. "The mass of people . . . seldom judge or determine right," as Alexander Hamilton put it during the framing of the Constitution, expressing a common elite view. Furthermore, the infamous three-fifths clause ensured the slave states an extra boost, a very significant issue considering their prominent role in the political and economic institutions. As the party system took shape in the nineteenth century, the Electoral College became a mirror of the state votes, which can give a result quite different from the popular vote because of the first-past-the-post rule—as it did once again in this election. Eliminating the Electoral College would be a good idea, but it's virtually impossible as the political system is now constituted. It is only one of many factors that contribute to the regressive character of the American political system, which, as Seth Ackerman observes in an interesting article in *Jacobin* magazine, would not pass muster by European standards.

Originally published in *Truthout*, January 6, 2017

Ackerman focuses on one severe flaw in the US system: the dominance of organizations that are not genuine political parties with public participation but rather elite-run candidate-selection institutions, often described not unrealistically as the two factions of the single business party that dominates the political system. They have protected themselves from competition by many devices that bar genuine political parties that grow out of free association of participants, as would be the case in a properly functioning democracy. Beyond that there is the overwhelming role of concentrated private and corporate wealth, not just in the presidential campaigns, as has been well documented particularly by Thomas Ferguson, but also in Congress. A recent study by Ferguson, Paul Jorgensen, and Jie Chen reveals a remarkably close correlation between campaign expenditures and electoral outcomes in Congress over decades. And extensive work in academic political science—particularly by Martin Gilens, Benjamin Page, and Larry Bartlett—reveals that most of the population is effectively unrepresented in that their attitudes and opinions have little or no effect on decisions of the people they vote for, which are pretty much determined by the very top of the income-wealth scale. In light of such factors as these, the defects of the Electoral College, while real, are of lesser significance.

To what extent is this presidential election a watershed moment for Republicans and Democrats alike?

For the eight years of the Obama presidency, the Republican organization has hardly qualified as a political party. A more accurate description was given by the respected political analysts Thomas Mann and Norman Ornstein of the conservative American Enterprise Institute: the party became an "insurgent outlier—ideologically extreme; contemptuous of the inherited social and economic policy regime; scornful of compromise; unpersuaded by conventional understanding of facts, evidence and science; and dismissive of the legitimacy of its political opposition." Its guiding principle was, whatever Obama tries to do, we have to block it, but without providing some sensible alternative. The goal was to make the country ungovernable, so that the insurgency could take power. Its infantile antics on the Affordable Care Act are a good illustration: endless votes to repeal it in favor of—nothing. Meanwhile the party has become split between the wealthy and privileged "establishment," devoted to the interests of their class, and the popular base that was mobilized when the establishment commitments to wealth and privilege became so extreme that it would be impossible to garner votes by presenting them accurately. It was therefore necessary to mobilize sectors that had always existed, but not as an organized political force: a strange amalgam of Christian evangelicals—a huge sector of the American population—nativists, white supremacists, white working- and lower-middle-class victims of the neoliberal policies of the past generation, and others who are fearful and angry, cast aside in the neoliberal economy while they perceive their traditional culture as being under attack. In past primaries, the candidates who rose from the base—Michele Bachmann, Herman Cain, Rick Santorum, and the rest—were so extreme that they were anathema to the establishment, who were able to use their ample resources to rid themselves of the plague and choose their favored candidate. The difference in 2016 is that they were unable to do it.

Now the party faces the task of formulating policies other than "No." It must find a way to craft policies that will somehow pacify or marginalize the popular base while serving the real constituency of the establishment. It is from this sector that Trump is picking his close associates and cabinet members: not exactly coal miners, iron and steel workers, small-business owners, or representatives of the concerns and demands of his voting base.

Democrats have to face the fact that for forty years they have pretty much abandoned whatever commitment they had to working people. It's quite shocking that Democrats have drifted so far from their modern New Deal origins that workers are now voting for their class enemy, not for the party of FDR. A return to some form of social democracy should not be impossible, as indicated by the remarkable success of the Sanders campaign, which departed radically from the norm of elections effectively bought by wealth and corporate power. It is important to bear in mind that his "political revolution," while quite appropriate for the times, would not have much surprised Dwight Eisenhower, another indication of the shift to the right during the neoliberal years.

If the Democratic Party is going to be a constructive force, it will have to develop and commit itself credibly to programs that address the valid concerns of the kind of people who voted for Obama, attracted by his message of "hope and change," and, when disillusioned by the disappearance of hope and the lack of change, switched to the con man who declared that he will bring back what they have lost. It will be necessary to face honestly the malaise of much of the country, including people like those in the Louisiana bayous whom Arlie Hochschild studied with such sensitivity and insight, and surely including the former working-class constituency of the Democrats. The malaise is revealed in many ways, not least by the astonishing fact that mortality has increased in the country, something unknown in modern industrial democracies apart from catastrophic events. That's particularly true among middle-aged whites, mainly traceable it seems to what are sometimes called "diseases of despair" (opioids, alcohol, suicide, and so on). A statistical analysis reported by the *Economist* found that these health metrics correlate with a remarkable 43 percent of the Republican Party's gains over the Democrats in the 2016 election and remain significant and predictive even when controlling for race, education, age, gender, income, marital status, immigration, and employment. These are all signs of severe collapse of much of the society, particularly in rural and working-class areas. Furthermore, such initiatives have to be undertaken alongside of firm dedication to the rights and needs of those sectors of the population that have historically been denied rights and repressed, often in harsh and brutal ways.

No small task, but not beyond reach, if not by the Democrats, then by some political party replacing them, drawing from popular movements—and through the constant activism of these movements, quite apart from electoral politics. Beyond that, those who perceive, rightly in my view, that the whole social and political system needs radical change, even if we are to survive, have their work cut out for them too.

Trump's cabinet is being filled by financial and corporate bigwigs and military leaders. Such selections hardly reconcile with his pre-election promises to "drain the swamp," so what should we expect from this megalomaniac and phony populist insofar as the future of the Washington establishment is concerned and the future of American democracy itself?

In this respect—note the qualification—*Time* magazine put it fairly well (December 26, 2016): "While some supporters may balk, Trump's decision to embrace those who have wallowed in the Washington muck has spread a sense of relief among the capital's political class. 'It shows,' says one GOP consultant close to the president-elect's transition, 'that he's going to govern like a normal Republican.'"

There surely is some truth to this. Business and investors plainly think so. The stock market boomed right after the election, led by the financial companies that Trump denounced during his campaign, particularly the leading demon of his rhetoric, Goldman Sachs. According to *Bloomberg News*, "The firm's surging stock price," up 30 percent in the month after the election, "has been the largest driver behind the Dow Jones Industrial Average's climb toward 20,000." The stellar market performance of Goldman Sachs is based largely on Trump's reliance on the demon to run the economy, buttressed by the promised roll-back in regulations, setting the stage for the next financial crisis (and taxpayer bailout). Other big gainers are energy corporations, health insurers, and construction firms, all expecting huge profits from the administration's announced plans. These include a Paul Ryan–style fiscal program of tax cuts for the rich and corporations, increased military spending, turning the health system over even more to insurance companies with predictable consequences, taxpayer largesse for a privatized form of credit-based infrastructure development, and other "normal Republican" gifts to wealth and privilege at taxpayer expense. Rather plausibly, economist Larry Summers describes the fiscal program as "the most misguided set of tax changes in US history [which] will massively favor the top 1 percent of income earners, threaten an explosive rise in federal debt, complicate the tax code and do little if anything to spur growth."

But great news for those who matter.

There are, however, some losers in the corporate system. Since November 8, gun sales, which more than doubled under Obama, have been dropping sharply, perhaps because of lessened fears that the government will take away the assault rifles and other armaments we need to protect ourselves from the feds. Sales rose through the year as polls showed Clinton in the lead, but after the election, the *Financial Times* reported, "shares in gunmakers such as Smith & Wesson and Sturm Ruger plunged." By mid-December, "the two companies had fallen 24 per cent and 17 per cent since the election, respectively." But all is not lost for the industry. As a spokesman explains, "To put it in perspective, US consumer sales of firearms are greater than the rest of the world combined. It's a pretty big market."

Normal Republicans cheer Trump's choice for Office of Management and Budget, Mick Mulvaney, one of the most extreme fiscal hawks, though a problem does arise: How will a fiscal hawk manage a budget designed to massively escalate the deficit? In a post-fact world, maybe that doesn't matter.

Also cheering to "normal Republicans" is the choice of the radically antilabor Andy Puzder for secretary of labor, though here, too, a contradiction may lurk in the background. As the ultra-rich CEO of restaurant chains, he relies on the most easily exploited nonunion labor for the dirty work, typically immigrants, which doesn't comport well with the plans to deport them en masse. The same problem arises for the infrastructure programs; the private firms that are set to profit from these initiatives rely heavily on the same labor source, though perhaps that problem can be finessed by redesigning the "beautiful wall" so that it will only keep out Muslims.

Is this to say then that Trump will be a "normal" Republican as America's forty-fifth president?

In such respects as the ones mentioned above, Trump proved himself very quickly to be a normal Republican, if to the extremist side. But in other respects he may not be a normal Republican, if that means something like a mainstream establishment Republican—people like Mitt Romney, who Trump went out of his way to humiliate in his familiar style, just as he did McCain and others of this category. But it's not only his style that causes offense and concern. His actions as well.

Take just the two most significant issues that we face, the most significant that humans have ever faced in their brief history on earth, issues that bear on species survival: nuclear war and global warming. Shivers went up the spine of many "normal Republicans," as of others who care about the fate of the species, when Trump tweeted that "The United States must greatly strengthen and expand its nuclear capability until such time as the world comes to its senses regarding nukes." Expanding nuclear capability means casting to the winds the treaties that have sharply reduced nuclear arsenals and that sane analysts hope may reduce them much further, in fact to zero, as advocated by such normal Republicans as Henry Kissinger and

Reagan's secretary of state, George Shultz, and by Reagan in some of his moments. Concerns did not abate when Trump went on to tell the cohost of TV show *Morning Joe*, "Let it be an arms race. We will outmatch them at every pass." And it wasn't too comforting even when his White House team tried to explain that the Donald didn't say what he said.

Nor do concerns abate because Trump was presumably reacting to Putin's statement: "We need to strengthen the military potential of strategic nuclear forces, especially with missile complexes that can reliably penetrate any existing and prospective missile defense systems. We must carefully monitor any changes in the balance of power and in the political-military situation in the world, especially along Russian borders, and quickly adapt plans for neutralizing threats to our country."

Whatever one thinks of these words, they have a defensive cast, and, as Putin has stressed, they are in large part a reaction to the highly provocative installation of a missile defense system on Russia's border on the pretext of defense against nonexistent Iranian weapons. Trump's tweet intensifies fears about how he might react when crossed, for example, by unwillingness of some adversary to bow to his vaunted negotiating skills. If the past is any guide, he might, after all, find himself in a situation where he must decide within a few minutes whether to blow up the world.

The other crucial issue is environmental catastrophe. It cannot be stressed too strongly that Trump won two victories on November 8: the lesser one in the Electoral College, and the greater one in Marrakech, where some two hundred countries were seeking to put teeth in the promises of the Paris negotiations on climate change. On Election Day, the conference heard a dire report on the state of the Anthropocene from the World Meteorological Organization. As the results of the election came in, the stunned participants virtually abandoned the proceedings, wondering if anything could survive the withdrawal of the most powerful state in world history. Nor can one stress too often the astonishing spectacle of the world placing its hopes for salvation in China, while the leader of the free world stands alone as a wrecking machine.

Although—amazingly—most ignored these astounding events, establishment circles did have some response. In *Foreign Affairs,* Varun Sivaram and Sagatom Saha warned of the costs to the United States of "ceding climate leadership to China," and the dangers to the world because China "would lead on climate-change issues only insofar as doing so would advance its national interests"—unlike the altruistic United States, which labors selflessly only for the benefit of mankind.

How intent Trump is on driving the world to the precipice was revealed by his appointments, including two militant climate change deniers, Myron Ebell and Scott Pruitt, to take charge of dismantling the Environmental Protection Agency that was established under Richard Nixon, with another denier slated to head the Department of the Interior.

But that's only the beginning. The cabinet appointments would be comical if the implications were not so serious. For Department of Energy, a man who said it should be eliminated (when he could remember its name) and is perhaps unaware that its main concern is nuclear weapons. For Department of Education, another billionaire, Betsy DeVos, who is dedicated to undermining and perhaps eliminating the public school system and who, as Lawrence Krause reminds us in the *New Yorker*, is a fundamentalist Christian member of a Protestant denomination holding that "all scientific theories be subject to Scripture" and that "Humanity is created in the image of God; all theorizing that minimizes this fact and all theories of evolution that deny the creative activity of God are rejected." Perhaps the department should request funding from Saudi sponsors of Wahhabi madrassas to help the process along.

DeVos's appointment is no doubt attractive to the evangelicals who flocked to Trump's standard and constitute a large part of the base of today's Republican Party. She should also be able to work amicably with vice president Mike Pence, one of the "prized warriors [of] a cabal of vicious zealots who have long craved an extremist Christian theocracy," as Jeremy Scahill details in the *Intercept*, reviewing his shocking record on other matters as well.

And so it continues, case by case. But not to worry. As James Madison assured his colleagues as they were framing the Constitution, a national republic would "extract from the mass of the Society the purest and noblest characters which it contains."

What about the choice of Rex Tillerson as secretary of state?

One partial exception to the above is choice of ExxonMobil CEO Rex Tillerson for secretary of state, which has aroused some hope among those across the spectrum who are rightly concerned with the rising and extremely hazardous tensions with Russia. Tillerson, like Trump in some of his pronouncements, has called for diplomacy rather than confrontation, which is all to the good—until we remember the sable lining of the beam of sunshine. The motive is to allow ExxonMobil to exploit vast Siberian oil fields and so to accelerate the race to disaster to which Trump and associates, and the Republican Party rather generally, are committed.

And how about Trump's national security staff—do they fit the mold of "normal" Republicans, or are they also part of the extreme right?

Normal Republicans might be somewhat ambivalent about Trump's national security staff. It is led by national security adviser General Michael Flynn, a radical Islamophobe who declares that Islam is not a religion but rather a political ideology, like fascism, which is at war with us so we must defend ourselves, presumably against the whole Muslim world—a fine recipe for generating terrorists, not to speak of far worse consequences. Like the Red menace of earlier years, this Islamic ideology is penetrating deep into American society, Flynn declaims. They are, naturally, being helped by Democrats, who have voted to impose Sharia law in Florida, much as their predecessors served the Commies, as Joe McCarthy famously demonstrated. Indeed, there are "over 100 cases around the country," including Texas, Flynn warned in a speech in San Antonio. To ward off the imminent threat, Flynn is a board member of ACT!, which pushes state laws banning Sharia law, plainly an imminent threat in states like Oklahoma, where 70 percent of voters approved legislation to prevent the courts from applying this grim menace to the judicial system.

Second to Flynn in the national security apparatus is secretary of defense General James "Mad Dog" Mattis, considered a relative moderate. Mad Dog has explained that "It's fun to shoot some people." He achieved his fame by leading the assault on Fallujah in November 2004, one of the most vicious crimes of the Iraq invasion. A man who is "just great," according to the president-elect: "the closest thing we have to Gen. George Patton."

In your view, is Trump bent on a collision course with China?

It's hard to say. Concerns were voiced about Trump's attitudes toward China, again full of contradictions, particularly his pronouncements on trade, which are almost meaningless in the current system of corporate globalization and complex international supply chains. Eyebrows were raised over his sharp departure from long-standing policy in his phone call with Taiwan's president, but even more by his implying that the United States might reject China's concerns over Taiwan unless China accepts his trade proposals, thus linking trade policy "to an issue of great-power politics over which China may be willing to go to war," the business press warned.

And what of Trump's views and stance on the Middle East? They seem to be in line with those of "normal" Republicans, right?

Unlike with China, normal Republicans did not seem dismayed by Trump's tweet foray into Middle East diplomacy, again breaking with standard protocol, demanding that Obama veto UN Security Council Resolution 2334, which reaffirmed

> that the policy and practices of Israel in establishing settlements in the Palestinian and other Arab territories occupied since 1967 have no legal validity and constitute a serious obstruction to achieving a comprehensive, just and lasting peace in the Middle East [and] *Calls once more upon* Israel, as the occupying Power, to abide scrupulously by the 1949 Fourth Geneva Convention, to rescind its previous measures and to desist from taking any action which would result in changing the legal status and geographical nature and materially affecting the demographic composition of the Arab territories occupied since 1967, including Jerusalem, and, in particular, not to transfer parts of its own civilian population into the occupied Arab territories. [Emphasis in original]

Nor did they object when he informed Israel that it can ignore the lame duck administration and just wait until January 20, when all will be in order. What kind of order? That remains to be seen. Trump's unpredictability serves as a word of caution.

What we know so far is Trump's enthusiasm for the religious ultra-right in Israel and the settler movement generally. Among his largest charitable contributions are gifts to the West Bank settlement of Beth El in honor of David Friedman, his choice as ambassador to Israel. Friedman is president of American Friends of Beth El Institutions. The settlement, which is at the religious ultranationalist extreme of the settler movement, is also a favorite of the family of Jared Kushner, Trump's son-in-law, reported to be one of Trump's closest advisers. A lead beneficiary of the Kushner family's contributions, the Israeli press reports, "is a yeshiva headed by a militant rabbi who has urged Israeli soldiers to disobey orders to evacuate settlements and who has argued that homosexual tendencies arise from eating certain foods." Other beneficiaries include "a radical yeshiva in Yitzhar that has served as a base for violent attacks against Palestinians' villages and Israeli security forces."

In isolation from the world, Friedman does not regard Israeli settlement activity as illegal and opposes a ban on construction for Jewish settlers in the West Bank and East Jerusalem. In fact, he appears to favor Israel's annexation of the West Bank. That would not pose a problem for the Jewish state, Friedman explains, since the number of Palestinians living in the West Bank is exaggerated, and therefore a large Jewish majority would remain after annexation. In a post-fact world, such pronouncements are legitimate, though they might become accurate in the boring world of fact after another mass expulsion. Jews who support the international consensus on a two-state settlement are not just wrong, Friedman explains. They are "worse than kapos," the Jews who were controlling other inmates in service to their Nazi masters in the concentration camps, the ultimate insult.

On receiving the report of his nomination, Friedman said he looked forward to moving the US embassy to "Israel's eternal capital, Jerusalem," in accord with Trump's announced plans. In the past, such proposals were withdrawn, but today they might actually be fulfilled, perhaps advancing the prospects of a war with the Muslim world, as Trump's national security adviser appears to recommend.

Returning to UNSC 2334 and its interesting aftermath, it is important to recognize that the resolution is nothing new. The quote given above was not from UNSC 2334 but from UNSC 446, March 12, 1979, reiterated in essence in 2334. UNSC 446 passed 12–0 with the United States abstaining, joined by the UK and Norway. Several resolutions followed reaffirming 446. One resolution of particular interest was even stronger than 446/2334, calling on Israel "to dismantle the existing settlements" (UNSC Resolution 465, March 1980). This resolution passed unanimously, no abstentions.

The government of Israel did not have to wait for the UN Security Council (and, more recently, the World Court) to learn that its settlements are in gross violation of international law. In September 1967, only weeks after Israel's conquest of the occupied territories, in a top-secret document the government was informed by the legal adviser to the Ministry of Foreign Affairs, the distinguished international lawyer Theodor Meron, that "civilian settlement in the administered territories [Israel's term for the occupied territories] contravenes explicit provisions of the Fourth Geneva Convention." Meron explained further that the prohibition against transfer of settlers to the occupied territories "is categorical and not conditional upon the motives for the transfer or its objectives. Its purpose is to prevent settlement in occupied territory of citizens of the occupying state." Meron therefore advised, "If it is decided to go ahead with Jewish settlement in the administered territories, it seems to me vital, therefore, that settlement is carried out by military and not civilian entities. It is also important, in my view, that such settlement is in the framework of camps and is, on the face of it, of a temporary rather than permanent nature."

Meron's advice was followed. Settlement has often been disguised by the subterfuge suggested, the "temporary military entities" turning out later to be civilian settlements. The device of military settlement also has the advantage of providing a means to expel Palestinians from their lands on the pretext that a military zone is being established. Deceit was scrupulously planned, beginning as soon as Meron's authoritative report was delivered to the government. As documented by Israeli scholar Avi Raz, in September 1967,

> on the day a second civilian settlement came into being in the West Bank, the government decided that "as a 'cover' for the purpose of [Israel's] diplomatic campaign" the new settlements should be presented as army settlements and the settlers should be given the necessary instructions in case they were asked about the nature of their settlement. The Foreign Ministry directed Israel's diplomatic missions to present the settlements in the occupied territories as military "strongpoints" and to emphasize their alleged security importance.

Similar practices continue to the present.

In response to the Security Council orders of 1979–80 to dismantle existing settlements and to establish no new ones, Israel undertook a rapid expansion of settlements with the cooperation of both of the major Israeli political blocs, Labor and Likud, always with lavish US material support.

The primary differences today are that the United States is now alone against the whole world, and that it is a different world. Israel's flagrant violations of Security Council orders, and of international law, are by now far more extreme than they were thirty-five years ago and are arousing far greater condemnation in much of the world. The contents of Resolutions 446 and 2334 are therefore taken more seriously. Hence the revealing reactions to 2334, and to secretary of state John Kerry's explanation of the US vote. In the Arab world, the reactions seem to have been muted: We've been here before. In Europe they were generally supportive. In the United States and Israel, in contrast, coverage and commentary were extensive, and there was considerable hysteria. These are further indications of the increasing isolation of the United States on the world stage. Under Obama, that is. Under Trump US isolation will likely increase further, and indeed already did, even before he took office, as we have seen.

Why did Obama choose abstention at this juncture, that is, only a month or so before the end of his presidency?

Just why Obama chose abstention rather than veto is an open question: we do not have direct evidence. But there are some plausible guesses. There had been some ripples of surprise (and ridicule) after Obama's February 2011 veto of a UNSC resolution calling for implementation of official US policy, and he may have felt that it would be too much to repeat it if he is to salvage anything of his tattered legacy among sectors of the population that have some concern for international law and human rights. It is also worth remembering that among liberal Democrats, if not Congress, and particularly among the young, opinion about Israel-Palestine has been moving toward criticism of Israeli policies in recent years, so much so that "60% of Democrats support imposing sanctions or more serious action" in reaction to Israeli settlements, according to a December 2016 Brookings Institute poll. By now the core of support for Israeli policies in the United States has shifted to the far right, including the evangelical base of the Republican Party. Perhaps these were factors in Obama's decision, with his legacy in mind.

The 2016 abstention aroused furor in Israel and in the US Congress as well, both Republicans and leading Democrats, including proposals to defund the UN in retaliation for the world's crime. Israeli Prime Minister Netanyahu denounced Obama for his "underhanded, anti-Israel" actions. His office accused Obama of "colluding" behind the scenes with this "gang-up" by the Security Council, producing particles of "evidence" that hardly rise to the level of sick humor. A senior Israeli official added that the abstention "revealed the true face of the Obama administration," adding that "now we can understand what we have been dealing with for the past eight years."

Reality is rather different. Obama has in fact broken all records in support for Israel, both diplomatic and financial. The reality is described accurately by Middle East specialist of the *Financial Times*'s David Gardner:

> Mr Obama's personal dealings with Mr Netanyahu may often have been poisonous, but he has been the most pro-Israel of presidents: the most prodigal with military aid and reliable in wielding the US veto at the Security Council. . . . The election of Donald Trump has so far brought little more than turbo-frothed tweets to bear on this and other geopolitical knots. But the auguries are ominous. An irredentist government in Israel tilted towards the ultra-right is now joined by a national populist administration in Washington fire-breathing Islamophobia.

Public commentary on Obama's decision and Kerry's justification was split. Supporters generally agreed with Thomas Friedman that "Israel is clearly now on a path toward absorbing the West Bank's 2.8 million Palestinians . . . posing a demographic and democratic challenge." In a *New York Times* review of the state of the two-state solution defended by Obama-Kerry and threatened with extinction by Israeli policies, Max Fisher asks, "Are there other solutions?" He then turns to the possible alternatives, all of them "multiple versions of the so-called one-state solution" that poses a "demographic and democratic challenge": too many Arabs—perhaps soon a majority—in a "Jewish and democratic state."

In the conventional fashion, commentators assume that there are two alternatives: the two-state solution advocated by the world, or some version of the "one-state solution." Ignored consistently is a third alternative, the one that Israel has been implementing quite systematically since shortly after the 1967 war and that is now very clearly taking shape before our eyes: a Greater Israel, sooner or later incorporated into Israel proper, including a vastly expanded Jerusalem (already annexed in violation of Security Council orders) and any other territories that Israel finds valuable, while excluding areas of heavy Palestinian population concentration and slowly removing Palestinians within the areas scheduled for incorporation within Greater Israel. As in neocolonies generally, Palestinian elites will be able to enjoy Western standards in Ramallah, with "90 per cent of the population of the West Bank living in 165 separate 'islands,' ostensibly under the control of the [Palestinian Authority]" but actual Israeli control, as reported by Nathan Thrall, senior analyst with the International Crisis Group. Gaza will remain under crushing siege, separated from the West Bank in violation of the Oslo Accords.

The third alternative is another piece of the "reality" described by David Gardner.

In an interesting and revealing comment, Netanyahu denounced the "gang-up" of the world as proof of "old-world bias against Israel," a phrase reminiscent of Donald Rumsfeld's Old Europe–New Europe distinction in 2003.

It will be recalled that the states of Old Europe were the bad guys, the major states of Europe, which dared to respect the opinions of the overwhelming majority of their populations and thus refused to join the United States in the crime of the century, the invasion of Iraq. The states of New Europe were the good guys, which overruled an even larger majority and obeyed the master. The most honorable of the good guys was Spain's José María Aznar, who rejected virtually unanimous opposition to the war in Spain and was rewarded by being invited to join Bush and Blair in announcing the invasion.

This quite illuminating display of utter contempt for democracy, along with others like it at the same time, passed virtually unnoticed, understandably. The task at the time was to praise Washington for its passionate dedication to democracy, as illustrated by "democracy promotion" in Iraq, which suddenly became the party line after the "single question" (will Saddam give up his WMD?) was answered the wrong way.

Netanyahu is adopting much the same stance. The old world that is biased against Israel is the entire UN Security Council—more specifically, anyone in the world who has some lingering commitment to international law and human rights. Luckily for the Israeli far right, that excludes the US Congress and—very forcefully—the president-elect and his associates.

The Israeli government is of course cognizant of these developments. It is therefore seeking to shift its base of support to authoritarian states such as Singapore, China, and Modi's right-wing Hindu nationalist India, now becoming a very natural ally with its drift toward ultranationalism, reactionary internal policies, and hatred of Islam. The reasons for Israel's looking in this direction for support are outlined by Mark Heller, principal research associate at Tel Aviv's Institution for National Security Studies. "Over the long term," he explains, "there are problems for Israel in its relations with western Europe and with the U.S.," while in contrast, the important Asian countries "don't seem to indicate much interest about how Israel gets along with the Palestinians, Arabs, or anyone else." In short, China, India, Singapore, and other favored allies are less influenced by the kinds of liberal and humane concerns that pose increasing threats to Israel.

The tendencies developing in world order merit some attention. As noted, the United States is becoming even more isolated than it has been in recent years, when US-run polls—unreported in the United States but surely known in Washington—revealed that world opinion regarded the United States as by far the leading threat to world peace, no one else even close. Under Obama, the United States is now alone in abstention on the illegal Israel settlements, against an otherwise unanimous Security Council. With President Trump joining his bipartisan congressional supporters on this issue, the United States will be even more isolated in the world in support of Israeli crimes. Since November 8, the United States is isolated on the much more crucial matter of global warming, a threat to the survival of organized human life in anything like its present form. If Trump makes good on his promise to exit from the Iran deal, it is likely that the other participants will persist, leaving the United States still more isolated from Europe. The United States is also much more isolated from its Latin American "backyard" than in the past and will be even more isolated if Trump backs off from Obama's halting steps to normalize relations with Cuba, undertaken to ward off the likelihood that the United States would be pretty much excluded from hemispheric organizations because of its continuing assault on Cuba, in international isolation.

Much the same is happening in Asia, as even close US allies (apart from

Japan), even the UK, flock to the China-based Asian Infrastructure Development Bank and the China-based Regional Comprehensive Economic Partnership (in this case, including Japan). The China-based Shanghai Cooperation Organization (SCO) incorporates the Central Asian states: Siberia, with its rich resources; India; Pakistan; and soon, probably Iran; and perhaps Turkey. The SCO has rejected the US request for observer status and demanded that the United States remove all military bases from the region.

Immediately after the Trump election, we witnessed the intriguing spectacle of German chancellor Angela Merkel taking the lead in lecturing Washington on liberal values and human rights. Meanwhile, since November 8, the world looks to China for leadership in saving the world from environmental catastrophe, while the United States, in splendid isolation once again, devotes itself to undermining these efforts.

US isolation is not complete, of course. As was made very clear in the reaction to Trump's electoral victory, the United States has the enthusiastic support of the xenophobic ultra-right in Europe, including its neofascist elements. The return of the right in parts of Latin America offers the United States opportunities for alliances there as well. And the United States retains its close alliance with the dictatorship of the Gulf and Egypt, and with Israel, which is also separating itself from more liberal and democratic sectors in Europe and linking with authoritarian regimes that are not concerned with Israel's violations of international law and harsh attacks on elementary human rights.

The developing picture suggests the emergence of a New World Order, one that is rather different from the usual portrayals within the doctrinal system.

The Republican Base Is "Out of Control"

C. J. POLYCHRONIOU: Noam, perhaps because more outrageous political characters are drawn into US politics than at any other time in the recent past, we have become witnesses of some strange developments, such as GOP candidates attacking "free trade" agreements and even someone like Donald Trump having turned against his fellow billionaires. Are we witnessing the end of the old economic establishment in American politics?

NOAM CHOMSKY: There is something new in the 2016 election, but it is not the appearance of candidates who frighten the old establishment. That has been happening regularly. It traces back to the shift of both parties to the right during the neoliberal years, the Republicans so far to the right that they are unable to get votes with their actual policies: dedication to the welfare of the very rich and the corporate sector. The Republican leadership has accordingly been compelled to mobilize a popular base on issues that are peripheral to their core concerns: the Second Coming, "open carry" in schools, Obama as a Muslim, lashing out at the weak and victimized, and the rest of the familiar fare. The base that they've put together has regularly produced candidates unacceptable to the establishment: Bachmann, Cain, Santorum, Huckabee. But the establishment has always been able to beat them down in the usual ways and get their own man (Mitt Romney). What

Originally published in *Truthout*, March 29, 2016

is different this time is that the base is out of control, and the establishment is almost going berserk.

Analogies should not be pressed too far, but the phenomenon is not unfamiliar. The German industrialists and financiers were happy to use the Nazis as a weapon against the working class and the left, assuming that they could be kept under control. Didn't quite work out that way.

All of this aside, the United States is not immune to the general decline of the mainstream political parties of the West, and the growth of political insurgencies on the right and left (though "left" means moderate social democracy, in practice)—one of the predictable consequences of the neoliberal policies that have undermined democracy and caused substantial harm to most of the population, the less privileged sectors. All familiar.

It appears that big-ticket conservative donors, like the Koch brothers, are turning their back on the Republican Party. If this is actually true, what might possibly be the explanation for this development?

The reason, I think, is that they are having a problem controlling the base they have mobilized, and are seeking some way to avoid a serious blow to their interests. It wouldn't entirely surprise me if they manage somehow to control the Republican National Convention and possibly even bring in someone like Paul Ryan. Not a prospect to welcome, in my opinion.

Stories about wealthy individuals financing politicians are as old as the country itself. So, in what ways has money reshaped American politics in our own era?

Nothing that is completely new. The standard scholarly work on this topic—Thomas Ferguson's outstanding studies in his book *Golden Rule* and more recent publications—traces the practices and the consequences back to the late nineteenth century, with particularly interesting results on the New Deal years, continuing to the present.

There are always new twists. One, which Ferguson has discussed, dates to Newt Gingrich's machinations in the 1990s. Prestigious and influential positions in Congress used to be granted on the basis of seniority and perceived achievement. Now, they are basically bought, which drives congressional representatives even deeper into the pockets of the rich. And Supreme Court decisions have accelerated the process.

In the past, the candidate with the most money won almost all the time. But Donald Trump seems to have changed the rules about politics in money as he has actually spent less money than his rivals. Has the power of money suddenly shrunk in an election year dominated by extreme voices?

Don't know the exact figures, but Trump seems to be putting plenty of money into the campaign. However, it is striking how huge money chests have failed. Jeb Bush is the clearest case. There is a very interesting article by Andrew Cockburn about this in the April 2016 issue of *Harper's*, reviewing studies that show that an enormous amount of the money poured into political campaigns with TV ads, and the like, serves primarily to enrich the networks and the professional consultants but with little effect on voting.[1] In contrast, face-to-face contact and direct canvasing, which are inexpensive—but require a lot of often volunteer labor—do have a measurable impact. Note that a separate matter is the question of the influence of the campaign spending by wealth and power on policy decisions, the kind of question that Ferguson has investigated.

What specific economic interests would you say are best represented by GOP candidates in the 2016 election?

The super-rich and the corporate sector, even more so than usual.

One of the great myths in American political culture revolves around "free-market" capitalism. The US economy is not a "free-market" economy, as most libertarians would point out, but the question is whether there can be a system of "free-market" capitalism, let alone whether it would be desirable to have one.

There have been examples of something like free-market capitalism. The distinguished economic historian Paul Bairoch points out that "there is no doubt that the Third World's compulsory economic liberalism in the nineteenth century is a major element in explaining the delay in its industrialization," or even "deindustrialization." There are many well-studied illustrations. Meanwhile, Europe and the regions that managed to stay free of its control developed, as Europe itself did, by radical violation of these principles. England and the United States are prime examples, as is the one area of the global South that resisted colonization and developed: Japan.

Like many other economic historians, Bairoch concludes from a broad survey that "it is difficult to find another case where the facts so contradict a dominant theory" as the doctrine that free markets were the engine of growth, a harsh lesson that the global South has learned over the years, again in the recent neoliberal period. There are classic studies of some of the inherent problems in "free market" development, like Karl Polyani's *The Great Transformation*, Rajani Kanth's *Political Economy and Laissez-Faire*, and a substantial literature in economic history and history of technology.

There are also fundamental problems of unregulated markets, such as the restriction of choice that they impose (excluding public goods, like mass transportation) and their ignoring of externalities, which by now spells virtual doom to the species.

A recent poll showed that more than nine in ten Americans said they would vote for a qualified presidential candidate who is Catholic, a woman, Black, Hispanic, or Jewish, but less than half said they would vote for a candidate who is a socialist. Why is socialism still a taboo in this country (although one must admit that socialism seems to be dead virtually everywhere else today in the Western world)?

A difficult question to discuss, because the word "socialism" (like most terms of political discourse) has been so vulgarized and politicized that it is not very useful. The essence of traditional socialism was workers' control over production, along with popular democratic control of other components of social, economic, and political life. There was hardly a society in the world more remote from socialism than Soviet Russia, which is presented as the leading "socialist" society. If that's what "socialism" is, then we ought to oppose it. In other uses, the post office, national health programs, and others are called "socialist," but they are not opposed by the public—including national health, supported, often by large majorities, for many years in the United States, and still today. The term "socialist" became taboo for reasons of Cold War ideology, which divorced the term from any useful meaning.

There are significant elements of something like authentic socialism in the Western world, notably worker-owned (and sometimes managed) enterprises, cooperatives with real participation, and much else. I think they can be thought of in Bakunin's terms, as creating the institutions of a more free and just society within the present one.

These days the United States seems to have a comparative advantage over other "developed" countries around the world only in military technology. In fact, the United States is beginning to resemble more and more a "third world" country, at least with regard to its infrastructure and the extent of the poverty and homelessness among a significant and constantly rising portion of the population. In your view, what factors have led to this dreadful state of affairs in what still remains a very rich country?

The United States is, to an unusual extent, a business-run society, without roots in traditional societies in which, with all their severe flaws, people had some kind of place. Its history as a settler-colonial and slave society has left its social and cultural legacy, along with other factors, such as the unusual role of religious fundamentalism. There have been large-scale, radical democratic movements in American history, like the agrarian populist and militant labor movements, but they were mostly crushed, often with considerable violence.

One consequence is what Walter Dean Burnham calls a "crucial comparative peculiarity of the American political system: the total absence of a socialist or laborite mass party as an organized competitor in the electoral market." He showed that this accounts for much of the "class-skewed abstention rates" that he demonstrated for the United States, and the downplaying of class-related issues in the largely business-run political system. In some ways the system is a legacy of the Civil War, which has never really been overcome. Today's "red states" are solidly based in the Confederacy, which was solidly Democratic before the civil rights movement and Nixon's "Southern strategy" shifted party labels.

In many ways the United States is a very free society—also in social practices, such as lack of the kind of relations of deference that one often finds elsewhere. But one consequence of the complex amalgam is the sad state of social justice. Although an extremely rich society, with incomparable advantages, the United States ranks very low in measures of social justice among the richer Organization for Economic Cooperation and Development (OECD) societies, alongside of Turkey, Mexico, and Greece. Infrastructure is a disaster. One can take a high-speed train in other developed societies, or from China to Kazakhstan, but not from Boston to Washington—maybe the most traveled corridor—where there hasn't been much of an improvement since I took the train sixty-five years ago.

Traditional Marxists speak of human society as consisting of two parts: base and superstructure. Would you say that the base dictates the superstructure in US society?

Don't have much to say. I don't find the framework particularly useful. Who holds dominant decision-making power in US society is not very obscure at a general level: concentrated economic power, mostly in the corporate system. When we look more closely, it is of course more complex, and the population is by no means powerless when it is organized and dedicated and liberated from illusions.

2016 Election Puts United States at Risk of "Utter Disaster"

C. J. POLYCHRONIOU: Noam, let's start with a reflective look at how the US 2016 presidential elections shape up in terms of the state of the country and its role in global affairs and the ideological viewpoints expressed by some of the leading candidates of both parties.

NOAM CHOMSKY: It cannot be overlooked that we have arrived at a unique moment in human history. For the first time, decisions have to be made right now that will literally determine the prospects for decent human survival, and not in the distant future. We have already made that decision for a huge number of species. Species destruction is at the level of 65 million years ago, the fifth extinction, ending the age of the dinosaurs. That also opened the way for small mammals, ultimately us, a species with unique capacities, including unfortunately the capacity for cold and savage destruction.

The nineteenth-century reactionary opponent of the Enlightenment, Joseph de Maistre, criticized Thomas Hobbes for adopting the Roman phrase, "Man is a wolf to man," observing that it is unfair to wolves, who do not kill for pleasure. The capacity extends to self-destruction, as we are now witnessing. It is presumed that the fifth extinction was caused by a huge asteroid that hit the earth. Now we are the asteroid. The impact on humans is already significant and will soon become incomparably worse unless decisive action is taken right now. Furthermore, the risk of nuclear war, always a grim shadow,

Originally published in *Truthout*, March 9, 2016

is increasing. That would end any further discussion. We may recall Einstein's response to a question about the weapons that would be used in the next war. He said that he didn't know, but the war after that would be fought with stone axes. Inspection of the shocking record reveals that it's a near miracle that disaster has been avoided this far, and miracles do not go on forever. And that the risk is increasing is unfortunately all too evident.

Fortunately, these destructive and suicidal capacities of human nature are balanced by others. There is good reason to believe that such Enlightenment figures as David Hume and Adam Smith, and the anarchist activist-thinker Peter Kropotkin, were correct in regarding sympathy and mutual aid as core properties of human nature. We'll soon find out which characteristics are in the ascendant.

Turning to your question, we can ask how these awesome problems are being addressed in the quadrennial electoral extravaganza. The most striking fact is that they are barely being addressed at all, by either party.

There's no need to review the spectacle of the Republican primaries. Commentators can barely conceal their disgust and concern for what it tells us about the country and contemporary civilization. The candidates have, however, answered the crucial questions. They either deny global warming or insist that nothing should be done about it, demanding, in effect, that we race even more rapidly to the precipice. Insofar as they have detectable policies, they seem to be intent to escalate military confrontation and threats. For these reasons alone, the Republican organization—one hesitates to call it a political party in any traditional sense—poses a threat of a novel and truly horrifying kind to the human species and to the others who are "collateral damage" as higher intelligence proceeds on its suicidal course.

On the Democratic side, there is at least some recognition of the danger of environmental catastrophe, but precious little in the way of substantive policy proposals. On Obama's programs of upgrading the nuclear arsenal, or such critical matters as the rapid (and mutual) military buildup on Russia's borders, I haven't been able to find any clear positions.

In general, the ideological positions of the Republican candidates seem to be more of the usual: stuff the pockets of the rich and kick the rest in the face. The two Democratic candidates range from the New Deal style of Sanders's programs to the "New Democrat/moderate Republican" Clinton version, driven a bit to the left under the impact of the Sanders challenge. On international affairs, and the awesome tasks we face, it seems at best "more of the same."

In your view, what has led to Donald Trump's rise, and is he simply another case of those typical right-wing, populist characters who frequently surface in the course of history whenever nations face severe economic crises or are on a national decline?

Insofar as the United States is facing "national decline," it's largely self-inflicted. True, the United States could not possibly maintain the extraordinary hegemonic power of the early post–World War II period, but it remains the potentially richest country in the world, with incomparable advantages and security, and in the military dimension, virtually matches the rest of the world combined and is technologically far more advanced than any collection of rivals.

Trump's appeal seems based largely on perceptions of loss and fear. The neoliberal assault on the world's populations, almost always harmful to them, and often severely so, has not left the United States untouched, even though it has been somewhat more resilient than others. The majority of the population has endured stagnation or decline while extraordinary and ostentatious wealth has accumulated in very few pockets. The formal democratic system has suffered the usual consequences of neoliberal socio-economic policies, drifting toward plutocracy.

No need to review again the grim details—for example, the stagnation of real male wages for forty years and the fact that since the last crash some 90 percent of wealth created has found its way to 1 percent of the population. Or the fact that the majority of the population—those lower on the income scale—are effectively disenfranchised in that their representatives ignore their opinions and preferences, heeding the super-rich funders and power brokers.

In part, Trump supporters—predominantly, it seems, lower-middle class, working class, less educated—are reacting to the perception, largely accurate, that they have simply been left by the wayside. It's instructive to compare the current scene with the Great Depression. Objectively, conditions in the '30s were far worse, and, of course, the United States was a much poorer country then. Subjectively, however, conditions then were far better. Among working-class Americans, despite very high unemployment and suffering, there was a sense of hopefulness, a belief that we will somehow come out of this working together. It was fostered by the successes of militant labor activism, often interacting with lively left political parties and other organizations. A fairly sympathetic administration responded with constructive measures, though always constrained by the enormous power of Southern Democrats, who were willing to tolerate welfare state measures as long as the despised Black population was marginalized. Importantly, there was a feeling that the country was on the road to a better future. All of this is lacking today, not least because of the successes of the bitter attacks on labor organization that took off as soon as the war ended.

In addition, Trump draws substantial support from nativists and racists—it's worth remembering that the United States has been at the extreme, even beyond South Africa, in the strength of white supremacy, as comparative studies by George Frederickson convincingly showed. The United States has never really transcended the Civil War and the horrendous legacy of oppression of African Americans for five hundred years. There is also a long history of illusions about Anglo-Saxon purity, threatened by waves of immigrants (and freedom for Blacks, and indeed for women, no small matter among patriarchal sectors). Trump's predominantly white supporters can see that their image of a white-run (and, for many, male-run) society is dissolving before their eyes. It is also worth remembering that although the United States is unusually safe and secure, it is also perhaps the most frightened country in the world, another feature of the culture with a long history.

Such factors such as these mix in a dangerous brew. Just thinking back over recent years, in a book over a decade ago I quoted the distinguished scholar of German history Fritz Stern, writing in the establishment journal *Foreign Affairs*, on "the descent in Germany from decency to Nazi barbarism." He added, pointedly, "Today, I worry about the immediate future of the United States, the country that gave haven to German-speaking refugees in the 1930s," himself included. With implications for here and now that no careful reader could miss, Stern reviewed Hitler's demonic appeal to his "divine mission" as "Germany's savior" in a "pseudoreligious transfiguration of politics" adapted to "traditional Christian forms," ruling a government dedicated to "the basic principles" of the nation, with "Christianity as the foundation of our national morality and the family as the basis of national life." Further, Hitler's hostility toward the "liberal secular state," shared by much of the Protestant clergy, drove forward "a historic process in which resentment against a disenchanted secular world found deliverance in the ecstatic escape of unreason."

The contemporary resonance is unmistakable.

Such reasons to "worry about the future of the United States" have not been lacking since. We might recall, for example, the eloquent and poignant manifesto left by Joseph Stack when he crashed his small plane into an office building in Austin, Texas, hitting an IRS office, committing suicide. In it he traced his bitter life story as a worker who was doing everything according to the rules, and being crushed, step by step, by the corruption and brutality of the corporate system and the state authorities. He was speaking for many people like him. His manifesto was mostly ridiculed or ignored, but it should have been taken very seriously, along with many other clear signs of what has been taking place.

Nonetheless, Cruz and Rubio appear to me to be both far more dangerous than Trump. I see them as the real monsters, while Trump reminds me a bit of Silvio Berlusconi. Do you agree with any of these views?

I agree—and, as you know, the Trump-Berlusconi comparison is current in Europe. I would also add Paul Ryan to the list. He is portrayed as the deep thinker of the Republicans, the serious policy wonk, with spreadsheets and the other apparatus of the thoughtful analyst. The few attempts to analyze his programs, after dispensing with the magic that is regularly introduced, conclude that his actual policies are to virtually destroy every part of the federal government that serves the interests of the general population, while expanding the military and ensuring that the rich and the corporate sector will be well attended to—the core Republican ideology when the rhetorical trappings are drawn aside.

America's youth seems to be captivated by Bernie Sanders's message. Are you surprised by how well he is holding up?

I am surprised. I didn't anticipate the success of his campaign. It is, however, important to bear in mind that his policy proposals would not have surprised President Eisenhower, and that they are pretty much in tune with popular sentiments over a long period, often considerable majorities. For example, his much-maligned call for a national health care system of the kind familiar in similar societies is supported right now by about 60 percent of the population, a very high figure considering the fact that it is subject to constant condemnation and has very limited articulate advocacy. And that popular support goes far back. In the late Reagan years, about 70 percent of the population thought that there should be a constitutional guarantee of health care, and 40 percent thought there already was such a guarantee—meaning that it is such an obvious desideratum that it must be in this sacred document.

When Obama abandoned a public option without consideration, it was supported by almost two-thirds of the population. And there is every reason to believe that there would be enormous savings if the United States adopted the far more efficient national health care programs of other countries, which have about half the health care expenditures of the United States and generally better outcomes. The same is true of his proposals for higher taxes on the rich, free higher education, and other parts of his domestic programs, mostly reflecting New Deal commitments and similar to policy choices during the most successful growth periods of the post–World War II period.

Under what scenario can Sanders possibly win the Democratic nomination?

Evidently, it would require very substantial educational and organizational activities. But my own feeling, frankly, is that these should be directed substantially toward developing a popular movement that will not fade away after the election, but will join with others to form the kind of activist force that has been instrumental in initiating and carrying forward needed changes and reforms in the past.

Is America still a democracy and, if not, do elections really matter?

With all its flaws, America is still a very free and open society, by comparative standards. Elections surely matter. It would, in my opinion, be an utter disaster for the country, the world, and future generations if any of the viable Republican candidates were to reach the White House, and if they continue to control Congress. Consideration of the overwhelmingly important questions we discussed earlier suffices to reach that conclusion, and it's not all. For such reasons as those I alluded to earlier, American democracy, always limited, has been drifting substantially toward plutocracy. But these tendencies are not graven in stone. We enjoy an unusual legacy of freedom and rights left to us by predecessors who did not give up, often under far harsher conditions than we face now. And it provides ample opportunities for work that is badly needed, in many ways, in direct activism and pressures in support of significant policy choices, in building viable and effective community organizations, revitalizing the labor movement, and also in the political arena, from school boards to state legislatures and much more.

Trump in the White House

C. J. POLYCHRONIOU: Noam, the unthinkable has happened. In contrast to all forecasts, Donald Trump scored a decisive victory over Hillary Clinton, and the man that Michael Moore described as a "wretched, ignorant, dangerous part-time clown and full-time sociopath" will be the next president of the United States. In your view, what were the deciding factors that led American voters to produce the biggest upset in the history of US politics?

NOAM CHOMSKY: Before turning to this question, I think it is important to spend a few moments pondering just what happened on November 8, a date that might turn out to be one of the most important in human history, depending on how we react.

No exaggeration.

The most important news of November 8 was barely noted, a fact of some significance in itself.

On November 8, the World Meteorological Organization (WMO) delivered a report at the international conference on climate change in Morocco (COP22), which was called in order to carry forward the Paris agreement of COP21. The WMO reported that the past five years were the hottest on record. It reported rising sea levels, soon to increase as a result of the unexpectedly rapid melting of polar ice, most ominously the huge Antarctic glaciers. Already, Arctic sea ice over the past five years is 28 percent below the average of the previous twenty-nine years, not only raising

Originally published in *Truthout*, November 14, 2016

sea levels but also reducing the cooling effect of polar ice reflection of solar rays, thereby accelerating the grim effects of global warming. The WMO reported further that temperatures are approaching dangerously close to the goal established by COP21, along with other dire reports and forecasts.

Another event took place on November 8, which also may turn out to be of unusual historical significance for reasons that, once again, were barely noted.

On November 8, the most powerful country in world history, which will set its stamp on what comes next, had an election. The outcome placed total control of the government—executive, Congress, the Supreme Court—in the hands of the Republican Party, which has become the most dangerous organization in world history.

Apart from the last phrase, all of this is uncontroversial. The last phrase may seem outlandish, even outrageous. But is it? The facts suggest otherwise. The party is dedicated to racing as rapidly as possible to destruction of organized human life. There is no historical precedent for such a stand.

Is this an exaggeration? Consider what we have just been witnessing.

During the Republican primaries, every candidate denied that what is happening is happening—with the exception of the sensible moderates, like Jeb Bush, who said it's all uncertain, but we don't have to do anything because we're producing more natural gas, thanks to fracking. Or John Kasich, who agreed that global warming is taking place, but added that "we are going to burn [coal] in Ohio and we are not going to apologize for it."

The winning candidate, now the president-elect, calls for rapid increase in use of fossil fuels, including coal; dismantling of regulations; rejection of help to developing countries that are seeking to move to sustainable energy; and, in general, racing to the cliff as fast as possible.

Trump has already taken steps to dismantle the Environmental Protection Agency (EPA) by placing in charge of the EPA transition a notorious (and proud) climate change denier, Myron Ebell. Trump's top adviser on energy, billionaire oil executive Harold Hamm, announced his expectations, which were predictable: dismantling regulations, tax cuts for the industry (and the wealthy and corporate sector generally), more fossil fuel production, lifting Obama's temporary block on the Dakota Access Pipeline. The market reacted quickly. Shares in energy corporations boomed, including the world's largest coal miner, Peabody Energy, which had filed for bankruptcy, but after Trump's victory registered a 50 percent gain.

The effects of Republican denialism had already been felt. There had been hopes that the COP21 Paris agreement would lead to a verifiable treaty, but any such thoughts were abandoned because the Republican Congress would not accept any binding commitments, so what emerged was a voluntary agreement, evidently much weaker.

Effects may soon become even more vividly apparent than they already are. In Bangladesh alone, tens of millions are expected to have to flee from low-lying plains in coming years because of sea level rise and more severe weather, creating a migrant crisis that will make today's pale in significance. With considerable justice, Bangladesh's leading climate scientist says that "These migrants should have the right to move to the countries from which all these greenhouse gases are coming. Millions should be able to go to the United States." And to the other rich countries that have grown wealthy while bringing about a new geological era, the Anthropocene, marked by radical human transformation of the environment. These catastrophic consequences can only increase, not just in Bangladesh, but in all of South Asia as temperatures, already intolerable for the poor, inexorably rise and the Himalayan glaciers melt, threatening the entire water supply. Already in India, some 300 million people are reported to lack adequate drinking water. And the effects will reach far beyond.

It is hard to find words to capture the fact that humans are facing the most important question in their history—whether organized human life will survive in anything like the form we know—and are answering it by accelerating the race to disaster.

Similar observations hold for the other huge issue concerning human survival: the threat of nuclear destruction, which has been looming over our heads for seventy years and is now increasing.

It is no less difficult to find words to capture the utterly astonishing fact that in all of the massive coverage of the electoral extravaganza, none of this receives more than passing mention. At least I am at a loss to find appropriate words.

Turning finally to the question raised, to be precise, it appears that Clinton received a slight majority of the vote. The apparent decisive victory has to do with curious features of American politics: among other factors, the Electoral College residue of the founding of the country as an alliance of separate states; the winner-take-all system in each state; the arrangement of congressional districts (sometimes by gerrymandering) to provide greater weight to rural votes (in past elections, and probably this one too, Democrats have had a comfortable margin of victory in the popular vote for the House but hold a minority of seats); the very high rate of abstention (usually close to half in presidential elections, this one included). Of some significance for the future is the fact that in the age eighteen-to-twenty-five range, Clinton won handily, and Sanders had an even higher level of support. How much this matters depends on what kind of future humanity will face.

According to current information, Trump broke all records in the support he received from white voters, working class and lower middle class, particularly in the $50,000 to $90,000 income range, rural and suburban, primarily those without college education. These groups share the anger throughout the West at the centrist establishment, revealed as well in the unanticipated Brexit vote and the collapse of centrist parties in continental Europe. Many of the angry and disaffected are victims of the neoliberal policies of the past generation, the policies described in congressional testimony by Federal Reserve chair Alan Greenspan—"St. Alan," as he was called reverentially by the economics profession and other admirers until the miraculous economy he was supervising crashed in 2007–2008, threatening to bring the whole world economy down with it. As Greenspan explained during his glory days, his successes in economic management were based substantially on "growing worker insecurity." Intimidated working people would not ask for higher wages, benefits, and security but would be satisfied with the stagnating wages and reduced benefits that signal a healthy economy by neoliberal standards.

Working people, who have been the subjects of these experiments in economic theory, are not particularly happy about the outcome. They are not, for example, overjoyed at the fact that in 2007, at the peak of the neoliberal miracle, real wages for nonsupervisory workers were lower than they had been years earlier, or that real wages for male workers are about at 1960s levels while spectacular gains have gone to the pockets of a very few at the top, disproportionately a fraction of 1 percent. Not the result of market forces, achievement, or merit, but rather of definite policy decisions, matters reviewed carefully by economist Dean Baker in recently published work.[1]

The fate of the minimum wage illustrates what has been happening. Through the periods of high and egalitarian growth in the '50s and '60s, the minimum wage—which sets a floor for other wages—tracked productivity. That ended with the onset of neoliberal doctrine. Since then, the minimum wage has stagnated (in real value). Had it continued as before, it would probably be close to $20 per hour. Today, it is considered a political revolution to raise it to $15.

With all the talk of near-full employment today, labor force participation remains below the earlier norm. And for working people, there is a great difference between a steady job in manufacturing with union wages and benefits, as in earlier years, and a temporary job with little security in some service profession. Apart from wages, benefits, and security, there is a loss of dignity, of hope for the future, of a sense that this is a world in which I belong and play a worthwhile role.

The impact is captured well in Arlie Hochschild's sensitive and illuminating portrayal of a Trump stronghold in Louisiana, where she lived and worked for many years.[2] She uses the image of a line in which residents are standing, expecting to move forward steadily as they work hard and keep to all the conventional values. But their position in the line has stalled. Ahead of them, they see people leaping forward, but that does not cause much distress, because it is "the American way" for (alleged) merit to be rewarded. What does cause real distress is what is happening behind them. They believe that "undeserving people" who do not "follow the rules" are being moved in front of them by federal government programs they erroneously see as designed to benefit African Americans, immigrants, and others they often regard with contempt. All of this is exacerbated by Ronald Reagan's racist fabrications about "welfare queens" (by implication Black) stealing white people's hard-earned money and other fantasies.

Sometimes failure to explain, itself a form of contempt, plays a role in fostering hatred of government. I once met a house painter in Boston who had turned bitterly against the "evil" government after a Washington bureaucrat who knew nothing about painting organized a meeting of painting contractors to inform them that they could no longer use lead paint—"the only kind that works," as they all knew, but the suit didn't understand. That destroyed his small business, compelling him to paint houses on his own with substandard stuff forced on him by government elites.

Sometimes there are also some real reasons for these attitudes toward government bureaucracies. Hochschild describes a man whose family and friends are suffering bitterly from the lethal effects of chemical pollution but who despises the government and the "liberal elites," because for him, the EPA means some ignorant guy who tells him he can't fish but does nothing about the chemical plants.

These are just samples of the real lives of Trump supporters, who are led to believe that Trump will do something to remedy their plight, though the merest look at his fiscal and other proposals demonstrates the opposite—posing a task for activists who hope to fend off the worst and to advance desperately needed changes.

Exit polls reveal that the passionate support for Trump was inspired primarily by the belief that he represented change, while Clinton was perceived as the candidate who would perpetuate their distress. The "change" that Trump is likely to bring will be harmful or worse, but it is understandable that the consequences are not clear to isolated people in an atomized society lacking the kinds of associations (like unions) that can educate and organize. That is a crucial difference between today's despair and the generally hopeful attitudes of many working people under much greater economic duress during the Great Depression of the 1930s.

There are other factors in Trump's success. Comparative studies show that doctrines of white supremacy have had an even more powerful grip on American culture than in South Africa, and it's no secret that the white population is declining. In a decade or two, whites are projected to be a minority of the work force, and not too much later, a minority of the population. The traditional conservative culture is also perceived as under attack by the successes of identity politics, regarded as the province of elites who have only contempt for the "hard-working, patriotic, church-going [white] Americans with real family values'" who see their familiar country as disappearing before their eyes.

One of the difficulties in raising public concern over the very severe threats of global warming is that 40 percent of the US population does not see why it is a problem, since Christ is returning in a few decades. About the same percentage believe that the world was created a few thousand years ago. If science conflicts with the Bible, so much the worse for science. It would be hard to find an analogue in other societies.

The Democratic Party abandoned any real concern for working people by the 1970s, and they have therefore been drawn to the ranks of their bitter class enemies, who at least pretend to speak their language—Reagan's folksy style of making little jokes while eating jelly beans, George W. Bush's carefully cultivated image of a regular guy you could meet in a bar who loved to cut brush on the ranch in 100-degree heat and his probably faked mispronunciations (it's unlikely that he talked like that at Yale), and now Trump, who gives voice to people with legitimate grievances—people who have lost not just jobs but also a sense of personal self-worth—and who rails against the government that they perceive as having undermined their lives (not without reason).

One of the great achievements of the doctrinal system has been to divert anger from the corporate sector to the government that implements the programs that the corporate sector designs, such as the highly protectionist corporate/investor rights agreements that are uniformly misdescribed as "free trade agreements" in the media and commentary. With all its flaws, the government is, to some extent, under popular influence and control, unlike the corporate sector. It is highly advantageous for the business world to foster hatred for pointy-headed government bureaucrats and to drive out of people's minds the subversive idea that the government might become an instrument of popular will, a government of, by, and for the people.

Is Trump representing a new movement in American politics, or was the outcome of this election primarily a rejection of Hillary Clinton by voters who hate the Clintons and are fed up with "politics as usual"?

It's by no means new. Both political parties have moved to the right during the neoliberal period. Today's New Democrats are pretty much what used to be called "moderate Republicans." The Republicans have moved so far toward a dedication to the wealthy and the corporate sector that they cannot hope to get votes on their actual programs, and have turned to mobilizing sectors of the population that have always been there, but not as an organized coalitional political force: evangelicals, nativists, racists, and the victims of the current forms of globalization. This version of globalization is designed to set working people around the world in competition with one another while protecting the privileged. It is furthermore designed to undermine the legal and other measures that provided working people with some protection and with ways to influence decision-making in the closely linked public and private sectors, notably with effective labor unions. None of this is intrinsic to globalization; rather, it is a specific form of investor-friendly globalization, a mixture of protectionism, investor rights, and some limited provisions about authentic trade.

The consequences have been evident in recent Republican primaries. Every candidate that has emerged from the base has been so extreme that the Republican establishment had to use its ample resources to beat them down. The difference in 2016 is that the establishment failed, much to its chagrin, as we have seen.

Deservedly or not, Clinton represented the policies that were feared and hated, while Trump was seen as the symbol of "change"—change of what kind requires a careful look at his actual proposals, something largely missing in what reached the public. The campaign itself was remarkable in its avoidance of issues, and media commentary generally complied, keeping to the concept that true "objectivity" means reporting accurately what is "within the beltway" but not venturing beyond.

Trump said, following the outcome of the election, that he "will represent all Americans." How is he going to do that when the nation is so divided and he has already expressed deep hatred for many groups in the United States, including women and minorities? Do you see any resemblance between Brexit and Donald Trump's victory?

There are definite similarities to Brexit, and also to the rise of the ultranationalist far-right parties in Europe—whose leaders were quick to congratulate Trump on his victory, perceiving him as one of their own: Nigel Farage, Marine Le Pen, Viktor Orban, and others like them. And these developments are quite frightening. A look at the polls in Austria and Germany—*Austria and Germany*—cannot fail to evoke unpleasant memories for those familiar with the 1930s, even more so for those who watched directly, as I did as a child. I can still recall listening to Hitler's speeches, not understanding the words, though the tone and audience reaction were chilling enough. The first article that I remember writing was in February 1939, after the fall of Barcelona, on the seemingly inexorable spread of the fascist plague. And by strange coincidence, it was from Barcelona that my wife and I watched the results of the 2016 US presidential election unfold.

As to how Trump will handle what he has brought forth—not created, but brought forth—we cannot say. Perhaps his most striking characteristic is unpredictability. A lot will depend on the reactions of those appalled by his performance and the visions he has projected, such as they are.

Trump has no identifiable political ideology guiding his stance on economic, social, and political issues, yet there are clear authoritarian tendencies in his behavior. Therefore, do you find any validity behind the claims that Trump may represent the emergence of "fascism with a friendly face" in the United States?

For many years, I have been writing and speaking about the danger of the rise of an honest and charismatic ideologue in the United States, someone who could exploit the fear and anger that has long been boiling in much of the society, and who could direct it away from the actual agents of malaise to vulnerable targets. That could indeed lead to what sociologist Bertram Gross called "friendly fascism" in a perceptive study thirty-five years ago. But that requires an honest ideologue, a Hitler type, not someone whose only detectable ideology is Me. The dangers, however, have been real for many years, perhaps even more so in the light of the forces that Trump has unleashed.

With the Republicans in the White House, but also controlling both houses and the future shape of the Supreme Court, what will the United States look like for at least the next four years?

A good deal depends on his appointments and circle of advisers. Early indications are unattractive, to put it mildly.

The Supreme Court will be in the hands of reactionaries for many years, with predictable consequences. If Trump follows through on his Paul Ryan–style fiscal programs, there will be huge benefits for the very rich—estimated by the Tax Policy Center as a tax cut of over 14 percent for the top 0.1 percent and a substantial cut more generally at the upper end of the income scale, but with virtually no tax relief for others, who will also face major new burdens. The respected economics correspondent of the *Financial Times*, Martin Wolf, writes: "The tax proposals would shower huge benefits on already rich Americans such as Mr. Trump," while leaving others in the lurch, including, of course, his constituency. The immediate reaction of the business world reveals that Big Pharma, Wall Street, the military industry, energy industries, and other such wonderful institutions expect a very bright future.

One positive development might be the infrastructure program that Trump has promised while (along with much reporting and commentary) concealing the fact that it is essentially the Obama stimulus program that would have been of great benefit to the economy, and to the society generally, but was killed by the Republican Congress on the pretext that it would explode the deficit. While that charge was spurious at the time, given the very low interest rates, it holds in spades for Trump's program, now accompanied by radical tax cuts for the rich and corporate sector and increased Pentagon spending.

There is, however, an escape, provided by Dick Cheney when he explained to Bush's treasury secretary Paul O'Neill that "Reagan proved that deficits don't matter"—meaning deficits that we Republicans create in order to gain popular support, leaving it to someone else, preferably Democrats, to somehow clean up the mess. The technique might work, for a while at least.

There are also many questions about foreign policy consequences, mostly unanswered.

There is mutual admiration between Trump and Putin. How likely is it therefore that we may see a new era in US–Russia relations?

One hopeful prospect is that there might be reduction of the very dangerous and mounting tensions at the Russian border: note "the Russian border," not the Mexican border. Thereby lies a tale that we cannot go into here. It is also possible that Europe might distance itself from Trump's America, as already suggested by German chancellor Angela Merkel and other European leaders—and from the British voice of American power, after Brexit.

That might possibly lead to European efforts to defuse the tensions, and perhaps even efforts to move toward something like Mikhail Gorbachev's vision of an integrated Eurasian security system without military alliances, rejected by the United States in favor of NATO expansion, a vision revived recently by Putin, whether seriously or not, we do not know, since the gesture was dismissed.

Is US foreign policy under a Trump administration likely to be more or less militaristic than what we have seen under the Obama administration or even the George W. Bush administration?

I don't think one can answer with any confidence. Trump is too unpredictable. There are too many open questions. What we can say is that popular mobilization and activism, properly organized and conducted, can make a large difference.

And we should bear in mind that the stakes are very large.

Global Warming and the Future of Humanity

C. J. POLYCHRONIOU: A consensus seems to be emerging among scientists and even political and social analysts that global warming and climate change represent the greatest threat to the planet. Do you concur with this view, and why?

NOAM CHOMSKY: I agree with the conclusion of the experts who set the Doomsday Clock for the *Bulletin of Atomic Scientists*. They have moved the clock two minutes closer to midnight—three minutes to midnight—because of the increasing threats of nuclear war and global warming. That seems to me a credible judgment. Review of the record shows that it's a near miracle that we have survived the nuclear age. There have been repeated cases when nuclear war came ominously close, often a result of malfunctioning of early-warning systems and other accidents, sometimes as a result of highly adventurist acts of political leaders. It has been known for some time that a major nuclear war might lead to nuclear winter that would destroy the attacker as well as the target. And threats are now mounting, particularly at the Russian border, confirming the prediction of George Kennan and other prominent figures that NATO expansion, particularly the way it was undertaken, would prove to be a "tragic mistake," a "policy error of historic proportions."

As for climate change, it's by now widely accepted by the scientific

Originally published in *Truthout*, September 17, 2016

community that we have entered a new geological era, the Anthropocene, in which the Earth's climate is being radically modified by human action, creating a very different planet, one that may not be able to sustain organized human life in anything like a form we would want to tolerate. There is good reason to believe that we have already entered the Sixth Extinction, a period of destruction of species on a massive scale, comparable to the Fifth Extinction 65 million years ago, when three-quarters of the species on earth were destroyed, apparently by a huge asteroid. Atmospheric carbon dioxide is rising at a rate unprecedented in the geological record since 55 million years ago. There is concern—to quote a statement by 150 distinguished scientists—that "global warming, amplified by feedbacks from polar ice melt, methane release from permafrost, and extensive fires, may become irreversible," with catastrophic consequences for life on Earth, humans included—and not in the distant future. Sea level rise—and destruction of water resources as glaciers melt—alone may have horrendous human consequences.

Virtually all scientific studies point to increased temperatures since 1975, and a recent story in the *New York Times* confirms that decades-long warnings by scientists on global warming are no longer theoretical, as land ice melts and sea levels rise.[1] Yet, there are still people out there who not only question the widely accepted scientific view that current climate change is mostly caused by human activities but also cast a doubt on the reliability of surface temperatures. Do you think this is all politically driven, or also caused by ignorance and perhaps even fear of change?

It is an astonishing fact about the current era that in the most powerful country in world history, with a high level of education and privilege, one of the two political parties virtually denies the well-established facts about anthropogenic climate change. In the primary debates for the 2016 election, every single Republican candidate was a climate change denier, with one exception, John Kasich—the "rational moderate"—who said it may be happening but we shouldn't do anything about it. For a long time, the media have downplayed the issue. The euphoric reports on US fossil fuel production, energy independence, and so on, rarely even mention the fact that these triumphs accelerate the race to disaster. There are other factors too, but under these circumstances, it hardly seems surprising that a considerable part of the population either joins the deniers or regards the problem as not very significant.

In global surveys, Americans are more skeptical than other people around the world over climate change.[2] Why is that? And what does it tell us about American political culture?

The United States is to an unusual extent a business-run society, where short-term concerns of profit and market share displace rational planning. The United States is also unusual in the enormous scale of religious fundamentalism. The impact on understanding of the world is extraordinary. In national polls almost half of those surveyed have reported that they believe that God created humans in their present form ten thousand years ago (or less) and that man shares no common ancestor with the ape. There are similar beliefs about the Second Coming. Senator James Inhofe, who headed the Senate Committee on the Environment, speaks for many when he assures us that "God's still up there and there's a reason for this to happen," so it is sacrilegious for mere humans to interfere.

Recent data related to global emissions of heat-treating gases suggest that we may have left behind us the period of constantly increased emissions.[3] Is there room here for optimism about the future of the environment?

There is always room for Gramsci's "optimism of the will." There are still many options, but they are diminishing. Options range from simple initiatives that are easily undertaken like weatherizing homes (which could also create many jobs), to entirely new forms of energy, perhaps fusion, perhaps new means of exploiting solar energy outside the Earth's atmosphere (which has been seriously suggested), to methods of decarbonization that might, conceivably, even reverse some of the enormous damage already inflicted on the planet. And much else.

Given that change in human behavior happens slowly and that it will take many decades before the world economy makes a shift to new, clean(er) forms of energy, should we look toward a technological solution to climate change?

Anything feasible and potentially effective should be explored. There is little doubt that a significant part of any serious solution will require advances of technology, but that can only be part of the solution. Other major changes are necessary. Industrial production of meat makes a huge contribution to global warming. The entire socioeconomic system is based on production for profit and a growth imperative that cannot be sustained.

There are also fundamental issues of value: What is a decent life? Should the master-servant relation be tolerated? Should one's goals really be maximization of commodities—Veblen's "conspicuous consumption"? Surely there are higher and more fulfilling aspirations.

Many in the progressive and radical community, including the Union of Concerned Scientists (UCS), are quite skeptical and even opposed to so-called geoengineering solutions. Is this the flip side of the coin to climate change deniers?

That does not seem to me a fair assessment. UCS and others like them may be right or wrong, but they offer serious reasons. That is also true of the very small group of serious scientists who question the overwhelming consensus, but the mass climate denier movements—like the leadership of the Republican Party and those they represent—are a different phenomenon altogether. As for geoengineering, there have been serious general critiques that I think cannot be ignored, like Clive Hamilton's, along with many positive assessments. It is not a matter for subjective judgment based on guesswork and intuition. Rather, these are matters that have to be considered seriously, relying on the best scientific understanding available, without abandoning sensible precautionary principles.

What immediate but realistic and enforceable actions could or should be taken to tackle the climate change threat?

Rapid ending of use of fossil fuels, sharp increase in renewable energy, research into new options for sustainable energy, significant steps toward conservation, and, not least, a far-reaching critique of the capitalist model of human and resource exploitation; even apart from its ignoring of externalities, the latter is a virtual death knell for the species.

Is there a way to predict how the world will look like fifty years from now if humans fail to tackle and reverse global warming and climate change?

If current tendencies persist, the outcome will be disastrous before too long. Large parts of the world will become barely habitable, affecting hundreds of millions of people, along with other disasters that we can barely contemplate.

The Long History of US Meddling in Foreign Elections

C. J. POLYCHRONIOU: Noam, the US intelligence agencies have accused Russia of interference in the US presidential election in order to boost Trump's chances, and some leading Democrats have actually gone on record saying that the Kremlin's canny operatives changed the election outcome. What's your reaction to all this talk in Washington and among media pundits about Russian cyber- and propaganda efforts to influence the outcome of the presidential election in Donald Trump's favor?

NOAM CHOMSKY: Much of the world must be astonished—if they are not collapsing in laughter—while watching the performances in high places and in media concerning Russian efforts to influence an American election, a familiar US government specialty as far back as we choose to trace the practice. There is, however, merit in the claim that this case is different in character: by US standards, the Russian efforts are so meager as to barely elicit notice.

Let's talk about the long history of US meddling in foreign political affairs, which has always been morally and politically justified as the spread of American-style democracy throughout the world.

The history of US foreign policy, especially after World War II, is pretty

---

Originally published in *Truthout*, January 19, 2017. Some of the material for this interview was adapted from excerpts from *Deterring Democracy* (Verso Books, 1991).

much defined by the subversion and overthrow of foreign regimes, including parliamentary regimes, and the resort to violence to destroy popular organizations that might offer the majority of the population an opportunity to enter the political arena.

Following World War II, the United States was committed to restoring the traditional conservative order. To achieve this aim, it was necessary to destroy the antifascist resistance, often in favor of Nazi and fascist collaborators, to weaken unions and other popular organizations, and to block the threat of radical democracy and social reform, which were live options under the conditions of the time. These policies were pursued worldwide: in Asia, including South Korea, the Philippines, Thailand, Indochina, and, crucially, Japan; in Europe, including Greece, Italy, France, and crucially, Germany; in Latin America, including what the CIA took to be the most severe threats at the time, "radical nationalism" in Guatemala and Bolivia.

Sometimes the task required considerable brutality. In South Korea, about 100,000 people were killed in the late 1940s by security forces installed and directed by the United States. This was before the Korean War, which Jon Halliday and Bruce Cumings describe as "in essence" a phase—marked by massive outside intervention—in "a civil war fought between two domestic forces: a revolutionary nationalist movement, which had its roots in tough anti-colonial struggle, and a conservative movement tied to the status quo, especially to an unequal land system," restored to power under the US occupation. In Greece in the same years, hundreds of thousands were killed, tortured, imprisoned, or expelled in the course of a counterinsurgency operation, organized and directed by the United States, which restored traditional elites to power, including Nazi collaborators, and suppressed the peasant- and worker-based communist-led forces that had fought the Nazis. In the industrial societies, the same essential goals were realized, but by less violent means.

Yet it is true that there have been cases where the United States was directly involved in organizing coups even in advanced industrial democracies, such as in Australia and Italy in the mid-1970s. Correct?

Yes, there is evidence of CIA involvement in a virtual coup that overturned the Whitlam Labor government in Australia in 1975, when it was feared that Whitlam might interfere with Washington's military and intelligence bases in Australia. Large-scale CIA interference in Italian politics has been public knowledge since the congressional Pike Report was leaked in 1976, citing a figure of over $65 million to approved political parties and affiliates from 1948 through the early 1970s. In 1976, the Aldo Moro government fell in Italy after revelations that the CIA had spent $6 million to support anti-communist candidates. At the time, the European communist parties were moving toward independence of action with pluralistic and democratic tendencies (Eurocommunism), a development that in fact pleased neither Washington nor Moscow. For such reasons, both superpowers opposed the legalization of the Communist Party of Spain and the rising influence of the Communist Party in Italy, and both preferred center-right governments in France. Secretary of state Henry Kissinger described the "major problem" in the Western alliance as "the domestic evolution in many European countries," which might make Western communist parties more attractive to the public, nurturing moves toward independence and threatening the NATO alliance.

US interventions in the political affairs of other nations have always been morally and politically justified as part of the faith in the doctrine of spreading American-style democracy, but the actual reason was of course the spread of capitalism and the dominance of business rule. Was faith in the spread of democracy ever tenable?

No belief concerning US foreign policy is more deeply entrenched than the one regarding the spread of American-style democracy. The thesis is commonly not even expressed, merely presupposed as the basis for reasonable discourse on the US role in the world.

The faith in this doctrine may seem surprising. Nevertheless, there is a sense in which the conventional doctrine is tenable. If by "American-style democracy," we mean a political system with regular elections but no serious challenge to business rule, then US policy-makers doubtless yearn to see it established throughout the world. The doctrine is therefore not undermined by the fact that it is consistently violated under a different interpretation of the concept of democracy: as a system in which citizens may play some meaningful part in the management of public affairs.

So, what lessons can be drawn from all this about the concept of democracy as understood by US policy planners in their effort to create a new world order?

One problem that arose as areas were liberated from fascism after World War II was that traditional elites had been discredited, while prestige and influence had been gained by the resistance movement, based largely on groups responsive to the working class and poor, and often committed to some version of radical democracy. The basic quandary was articulated by Churchill's trusted adviser, South African prime minister Jan Christiaan Smuts, in 1943, with regard to Southern Europe: "With politics let loose among those peoples," he said, "we might have a wave of disorder and wholesale communism." Here the term "disorder" is understood as threat to the interests of the privileged, and "communism," in accordance with usual convention, refers to failure to interpret "democracy" as elite dominance, whatever the other commitments of the "communists" may be. With politics let loose, we face a "crisis of democracy," as privileged sectors have always understood.

In brief, at that moment in history, the United States faced the classic dilemma of third-world intervention in large parts of the industrial world as well. The US position was "politically weak" though militarily and economically strong. Tactical choices are determined by an assessment of strengths and weaknesses. The preference has, quite naturally, been for the arena of force and for measures of economic warfare and strangulation, where the United States has ruled supreme.

Wasn't the Marshall Plan a tool for consolidating capitalism and spreading business rule throughout Europe after World War II?

Very much so. For example, the extension of Marshall Plan aid in countries like France and Italy was strictly contingent on exclusion of communists—including major elements of the antifascist resistance and labor—from the government, "democracy," in the usual sense. US aid was critically important in early years for suffering people in Europe and was therefore a powerful lever of control, a matter of much significance for US business interests and longer-term planning. The fear in Washington was that the communist left would emerge victorious in Italy and France without massive financial assistance.

On the eve of the announcement of the Marshall Plan, ambassador to France Jefferson Caffery warned Secretary of State Marshall of grim consequences if the communists won the elections in France: "Soviet penetration of Western Europe, Africa, the Mediterranean, and the Middle East would be greatly facilitated" (May 12, 1947). The dominoes were ready to fall. During May, the United States pressured political leaders in France and Italy to form coalition governments excluding the communists. It was made clear and explicit that aid was contingent on preventing an open political competition, in which left and labor might dominate. Through 1948, Secretary of State Marshall and others publicly emphasized that if communists were voted into power, US aid would be terminated; no small threat, given the state of Europe at the time.

In France, the postwar destitution was exploited to undermine the French labor movement, along with direct violence. Desperately needed food supplies were withheld to coerce obedience, and gangsters were organized to provide goon squads and strike breakers, a matter that is described with some pride in semiofficial US labor histories, which praise the AFL (American Federation of Labor) for its achievements in helping to save Europe by splitting and weakening the labor movement (thus frustrating alleged Soviet designs) and safeguarding the flow of arms to Indochina for the French war of re-conquest, another prime goal of the US labor bureaucracy. The CIA reconstituted the Mafia for these purposes, in one of its early operations. The quid pro quo was restoration of the heroin trade. The US government connection to the drug boom continued for many decades.

US policies toward Italy basically picked up where they had been broken off by World War II. The United States had supported Mussolini's fascism from the 1922 takeover through the 1930s. Mussolini's wartime alliance with Hitler terminated these friendly relations, but they were reconstituted as US forces liberated southern Italy in 1943, establishing the rule of field marshall Pietro Badoglio and the royal family that had collaborated with the Fascist government. As Allied forces drove toward the north, they dispersed the antifascist resistance along with local governing bodies it had formed in its attempt to establish a new democratic state in the zones it had liberated from Germany. Eventually, a center-right government was established with neofascist participation and the left soon excluded.

Here too, the plan was for the working classes and the poor to bear the burden of reconstruction, with lowered wages and extensive firing. Aid was contingent on removing communists and left socialists from office, because they defended workers' interests and thus posed a barrier to the intended style of recovery, in the view of the State Department. The Communist Party was collaborationist; its position "fundamentally meant the subordination of all reforms to the liberation of Italy and effectively discouraged any attempt in northern areas to introduce irreversible political changes as well as changes in the ownership of the industrial companies . . . disavowing and discouraging those workers' groups that wanted to expropriate some factories," as Gianfranco Pasquino put it. But the party did try to defend jobs, wages, and living standards for the poor and thus "constituted a political and psychological barrier to a potential European recovery program," historian John Harper comments, reviewing the insistence of Kennan and others that communists be excluded from government though agreeing that it would be "desirable" to include representatives of what Harper calls "the democratic working class." The recovery, it was understood, was to be at the expense of the working class and the poor.

Because of its responsiveness to the needs of these social sectors, the Communist Party was labeled "extremist" and "undemocratic" by US propaganda, which also skillfully manipulated the alleged Soviet threat. Under US pressure, the Christian Democrats abandoned wartime promises about workplace democracy, and the police, sometimes under the control of ex-fascists, were encouraged to suppress labor activities. The Vatican announced that anyone who voted for the Communists in the 1948 election would be denied sacraments, and backed the conservative Christian Democrats under the slogan *O con Cristo o contro Cristo* (Either with Christ or against Christ). A year later, Pope Pius excommunicated all Italian Communists.

A combination of violence, manipulation of aid and other threats, and a huge propaganda campaign sufficed to determine the outcome of the critical 1948 election, essentially bought by US intervention and pressures.

The CIA operations to control the Italian elections, authorized by the National Security Council in December 1947, were the first major clandestine operation of the newly formed agency. CIA operations to subvert Italian democracy continued into the 1970s at a substantial scale.

In Italy, as well as elsewhere, US labor leaders, primarily from the AFL, played an active role in splitting and weakening the labor movement and inducing workers to accept austerity measures while employers reaped rich profits. In France, the AFL had broken dock strikes by importing Italian scab labor paid by US businesses. The State Department called on the federation's leadership to exercise their talents in union-busting in Italy as well, and they were happy to oblige. The business sector, formerly discredited by its association with Italian fascism, undertook a vigorous class war with renewed confidence. The end result was the subordination of the working class and the poor to the traditional rulers.

Later commentators tend to see the US subversion of democracy in France and Italy as a defense of democracy. In a highly regarded study of the CIA and American democracy, Rhodri Jeffreys-Jones describes "the CIA's Italian venture," along with its similar efforts in France, as "a democracy-propping operation," though he concedes that "the selection of Italy for special attention . . . was by no means a matter of democratic principle alone"; our passion for democracy was reinforced by the strategic importance of the country. But it was a commitment to "democratic principle" that inspired the US government to impose the social and political regimes of its choice, using the enormous power at its command and exploiting the privation and distress of the victims of the war, who must be taught not to raise their heads if we are to have true democracy.

A more nuanced position is taken by James Miller in his monograph on US policies toward Italy. Summarizing the record, he concludes:

> In retrospect, American involvement in the stabilization of Italy was a significant, if troubling, achievement. American power assured Italians the right to choose their future form of government and also was employed to ensure that they chose democracy. In defense of that democracy against real but probably overestimated foreign and domestic threats, the United States used undemocratic tactics that tended to undermine the legitimacy of the Italian state.

The "foreign threats," as he had already discussed, were hardly real; the Soviet Union watched from a distance as the United States subverted the 1948 election and restored the traditional conservative order, keeping to its wartime agreement with Churchill that left Italy in the Western zone. The "domestic threat" was the threat of democracy.

The idea that US intervention provided Italians with freedom of choice while ensuring that they chose "democracy" (in our special sense of the term) is reminiscent of the attitude of the extreme doves toward Latin America: that its people should choose freely and independently—as long as doing so did not impact US interests adversely.

The democratic ideal, at home and abroad, is simple and straightforward: you are free to do what you want, as long as it is what we want you to do.

The Legacy of the Obama Administration

C. J. POLYCHRONIOU: Barack Obama was elected in 2008 as president of the United States in a wave of optimism, but at a time when the country was in the full grip of the financial crisis brought about, according to Obama himself, by "the reckless behavior of a lot of financial institutions around the world" and "the folks on Wall Street." Obama's rise to power has been well documented, including the funding of his Illinois political career by the well-known Chicago real estate developer and power peddler Tony Rezko, but the legacy of his presidency has yet to be written. First, in your view, did Obama rescue the US economy from a meltdown, and, second, did he initiate policies to ensure that "reckless financial behavior" would be kept at bay?

NOAM CHOMSKY: On the first question, the matter is debated. Some economists argue that the bank rescues were not necessary to avoid a serious depression, and that the system would have recovered, probably with some of the big banks broken up. Dean Baker for one. I don't trust my own judgment enough to take a strong position.

On the second question, Dodd-Frank takes some steps forward—making the system more transparent, greater reserve requirements, and so on—but congressional intervention has cut back some of the regulation, for example, of derivative transactions, leading to strong protests of

Originally published in *Truthout*, June 2, 2016

Dodd-Frank. Some commentators, Matt Taibbi for one, have argued that the Wall Street–Congress conniving undermined much of the force of the reform from the start.

What do you think were the real factors behind the 2008 financial crisis?

The immediate cause of the crisis was the housing bubble, based substantially on very risky subprime mortgage loans along with exotic financial instruments devised to distribute risk, reaching such complexity that few understand who owes what to whom. The more fundamental reasons have to do with basic market inefficiencies. If you and I agree on some transaction (say, you sell me a car), we may make a good bargain for ourselves, but we do not take into account the effect on others (pollution, traffic congestion, increase in price of gas, and more). These externalities, so called, can be very large. In the case of financial institutions, the effect is to underprice risk by ignoring "systemic risk." Thus if Goldman Sachs lends money, it will, if well managed, take into account the potential risk to itself if the borrower cannot pay, but not the risk to the financial system as a whole. The result is that risk is underpriced. There is too much risk for a sound economy. That can, in principle, be controlled by sound regulation, but financialization of the economy has been accompanied by deregulation mania, based on theological notions of "efficient markets" and "rational choice." Interestingly enough, several of the people who had primary responsibility for these destructive policies were chosen as Obama's leading economic policy advisers (Robert Rubin, Larry Summers, Tim Geithner, and others) during his first term in the White House. Alan Greenspan, the great hero of a few years ago, eventually conceded quietly that he did not understand how markets work—which is quite remarkable.

There are also other devices that lead to underpricing risk. Government rules on corporate governance provide perverse incentives: CEOs are highly rewarded for taking short-term risks, and can leave the ruins to someone else, floating away on their "golden parachutes," when collapse comes. And there is much more.

Didn't the 2008 financial crisis reveal once again that capitalism is a

parasitic system?

It is worth bearing in mind that "really existing capitalism" is remote from capitalism—at least in the rich and powerful countries. Thus in the United States, the advanced economy relies crucially on the dynamic state sector to socialize cost and risk while privatizing eventual profit—and "eventual" can be a long time: in the case of the core of the modern high-tech economy, computers and the Internet, it was decades. There is much more mythology that has to be dismantled if the questions are to be seriously posed.

Existing state-capitalist economies are indeed "parasitic" on the public, in the manner indicated, and others: bailouts (which are very common, in the industrial system as well), highly protectionist "trade" measures that guarantee monopoly pricing rights to state-subsidized corporations, and many other devices.

During his first term as president, you admitted that Obama faced an exceptionally hostile crowd on Capitol Hill, which of course remained hostile throughout his two terms. Be that as it may, was Obama ever a real reformer or was he more of a public manipulator who used popular political rhetoric to sideline the progressive mood of the country in an era of great inequality and mass discontent over the future of the United States?

Obama had congressional support for his first two years in office, the time when most presidential initiatives are introduced. I never saw any indication that he intended substantive progressive steps. I wrote about him before the 2008 primaries, relying on the Web page in which he presented himself as a candidate. I was singularly unimpressed, to put it very mildly. Actually, I was shocked, for the reasons I discussed.

Consider what Obama and his supporters regard as his signature achievement, the Affordable Care Act. At first, a public option (effectively, national health care) was dangled. It had almost two-thirds popular support. It was dropped without apparent consideration. The outlandish legislation barring the government from negotiating drug prices was opposed by some 85 percent of the population, but was kept with little discussion. The act is an improvement on the existing international scandal, but not by much, and with fundamental flaws.

Consider nuclear weapons. Obama had some nice things to say—nice enough to win the Nobel Peace Prize. There has been some progress, but it has been slight, and current moves are in the wrong direction.

In general: much smooth rhetoric, some positive steps, some regression, overall not a very impressive record. That seems to me a fair assessment, even putting aside the quite extraordinary stance of the Republican Party, which made it clear right after Obama's election that they were, substantially, a one-issue party: prevent the president from doing anything, no matter what happens to the country and the world. It is difficult to find analogues among industrial democracies. Small wonder that the most respected conservative political analysts (such as Thomas Mann or Norman Ornstein of the conservative American Enterprise Institute) refer to the party as a "radical insurgency" that has abandoned normal parliamentary politics.

In the foreign policy realm, Obama claimed to strive for a new era in the United States, away from the militarism of his predecessor and toward respect for international law and active diplomacy. How would you judge US foreign and military strategy under the Obama administration?

He has been more reluctant to engage troops on the ground than some of his predecessors and advisers, and instead has rapidly escalated special operations and his global assassination (drone) campaign, a moral disaster and arguably illegal as well.[1] On other fronts, it is a mixed story. Obama has continued to bar a nuclear weapons–free (technically, WMD-free) zone in the Middle East, evidently motivated by the need to protect Israeli nuclear weapons from scrutiny. By so doing, he is endangering the Nonproliferation Treaty, the most important disarmament treaty, which is contingent on establishing such a zone. He is dangerously escalating tensions along the Russian border, extending earlier policies. His trillion-dollar program for modernizing the nuclear weapons system is the opposite of what should be done. The investor-rights agreements (called "free trade agreements") are likely to be generally harmful to populations and beneficial to the corporate sector. Sensibly, he bowed to strong hemispheric pressures and took steps toward normalization of relations with Cuba. These and other moves amount to a mixed story, ranging from criminal to moderate improvement.

Looking at the state of the US economy, one can easily argue that the effects of the financial crisis of 2007–2008 are not only still around but that we have in place a set of policies that continue to suppress the standard of living for the working population and produce immense economic insecurity. Is this because of neoliberalism and the peculiarities of the nature of the US economy, or are there global and systemic forces at play such as the free movement of capital, automation, and the end of industrialization?

The neoliberal assault on the population remains intact, though less so in the United States than in Europe. Automation is not a major factor, and industrialization isn't ending, just being offshored. Financialization has of course exploded during the neoliberal period, and the general policies, pretty much global in character, are designed to enhance private and corporate power. That sets off a vicious cycle in which concentration of wealth leads to concentration of political power, which in turn yields legislation and administrative practices that carry the process forward. There are countervailing forces, and they might become more powerful. The potential is there, as we can see from the Sanders campaign and even the Trump campaign, if the white working class to which Trump appeals can become organized to focus on their real interests instead of being in thrall to their class enemy.

To the extent that Trump's programs are coherent, they fall into the same general category of those of Paul Ryan, who has granted us the kindness of spelling them out: increase spending on the military (already more than half of discretionary spending and almost as much as the rest of the world combined), and cut back taxes, mainly on the rich, with no new revenue sources. In brief, nothing much is left for any government program that might be of benefit to the general population and the world. Trump produces so many arbitrary and often self-contradictory pronouncements that it isn't easy to attribute to him a program, but he regularly keeps within this range—which, incidentally, means that his claims about supporting Social Security and Medicare are worthless.

Since the white working class cannot be mobilized to support the class enemy on the basis of their actual programs, the "radical insurgency" called "the Republican Party" appeals to its constituency on what are called "social-cultural issues": religion, fear, racism, nationalism. The appeals are facilitated by the abandonment of the white working class by the Democratic Party, which offers them very little but "more of the same." It is then facile for the liberal professional classes to accuse the white working class of racism and other such sins, though a closer look often reveals that the manifestations of this deep-rooted sickness of the society are simply taking different forms among various sectors.

Obama's charisma and undoubtedly unique rhetorical skills were critical elements in his struggle to rise to power, while Donald Trump is an extrovert who seeks to project the image of a powerful personality who knows how to get things done even if he relies on the use of banalities to create the image he wants to create about himself as a future leader of a country. Do personalities really matter in politics, especially in our own era?

I am very much down on charismatic leaders, and as for strong ones, it depends on what they are working for. The best, in our own kind of societies, I think, are the FDR types, who react to, are sympathetic to, and encourage popular movements for significant reform. Sometimes, at least.

And politicians to be elected to a national office have to be pretty good actors, right?

Electoral campaigns, especially in the United States, are being run by the advertising industry. The 2008 political campaign of Barack Obama was voted by the advertising industry as the best marketing campaign of the year.

Obama's last State of the Union address had all the rhetoric of someone running for president, not someone who has been in office for more than seven years. What do you make of this—Obama's vision of how the country should be and function eight to ten years from now?

He spoke as if he had not been elected eight years ago. Obama had plenty of opportunities to change the course of the country. Even his "signature" achievement, the reform of the health care system, is a watered-down version, as I pointed out earlier. Despite the huge propaganda assault denouncing government involvement in health care, and the extremely limited articulate response, a majority of the population (and a huge majority of Democrats) still favor national health care, Obama didn't even try, even when he had congressional support.

You have argued that nuclear weapons and climate change represent the two biggest threats facing humankind. In your view, is climate change a direct effect of capitalism, the view taken by someone like Naomi Klein, or related to humanity and progress in general, a view embraced by the British philosopher John Gray?

Geologists divide planetary history into eras. The Pleistocene lasted millions of years, followed by the Holocene, which began at about the time of the agricultural revolution ten thousand years ago, and recently the Anthropocene, corresponding to the era of industrialization. What we call "capitalism"—in practice, various varieties of state-capitalism—tends in part to keep to market principles that ignore nonmarket factors in transactions: so-called externalities, the cost to Tom if Bill and Harry make a transaction. That is always a serious problem, like systemic risk in the financial system, in which case the taxpayer is called upon to patch up the "market failures." Another externality is destruction of the environment—but in this case the taxpayer cannot step in to restore the system. It's not a matter of "humanity and progress," but rather of a particular form of social and economic development, which need not be specifically capitalist; the authoritarian Russian statist (not socialist) system was even worse. There are important steps that can be taken within existing systems (carbon tax, alternative energy, conservation, and so on), and they should be pursued as much as possible, along with efforts to reconstruct society and culture to serve human needs rather than power and profit.

What do you think of certain geoengineering undertakings to clean up the environment, such as the use of carbon negative technologies to suck carbon from the air?

These undertakings have to be evaluated with great care, paying attention to issues ranging from narrowly technical ones to large-scale societal and environmental impacts that could be quite complex and poorly understood. Sucking carbon from the air is done all the time—planting forests—and can presumably be carried considerably further to good effect, but I don't have the special knowledge required to provide definite answers. Other more exotic proposals have to be considered on their own merits—and with due caution.

Some major oil-producing countries, such as Saudi Arabia, are in the process of diversifying their economies, apparently fully aware of the fact that the fossil fuel era will soon be over. In the light of this development, wouldn't US foreign policy toward the Middle East take a radically new turn once oil has ceased being the precious commodity that it has been up to now?

Saudi Arabian leaders are talking about this much too late. These plans should have been undertaken seriously decades ago. Saudi Arabia and the Gulf states may become uninhabitable in the not very distant future if current tendencies persist. In the bitterest of ironies, they have been surviving on the poison they produce that will destroy them—a comment that holds for all of us, even if less directly. How serious the plans are is not very clear. There are many skeptics. One Twitter comment is that they split the electricity ministry and the water ministries for fear of electrocution. That captures much of the general sentiment. It would be good to be surprised.

Socialism for the Rich, Capitalism for the Poor

C. J. POLYCHRONIOU: Noam, in several of your writings you question the usual view of the United States as an archetypical capitalist economy. Please explain.

NOAM CHOMSKY: Consider this: Every time there is a crisis, the taxpayer is called on to bail out the banks and the major financial institutions. If you had a real capitalist economy in place, that would not be happening. Capitalists who made risky investments and failed would be wiped out. But the rich and powerful do not want a capitalist system. They want to be able to run the nanny state so when they are in trouble the taxpayer will bail them out. The conventional phrase is "too big to fail."

The IMF did an interesting study a few years ago on profits of the big US banks. It attributed most of them to the many advantages that come from the implicit government insurance policy—not just the featured bailouts, but access to cheap credit and much else—including things the IMF researchers didn't consider, like the incentive to undertake risky transactions, hence highly profitable in the short term, and if anything goes wrong, there's always the taxpayer. *Bloomberg Businessweek* estimated the implicit taxpayer subsidy at over $80 billion per year.

Much has been said and written about economic inequality. Is economic

Originally published in *Truthout*, December 10, 2016

inequality in the contemporary capitalist era very different from what it was in other post-slavery periods of American history?

The inequality in the contemporary period is almost unprecedented. If you look at total inequality, it ranks among the worse periods of American history. However, if you look at inequality more closely, you see that it comes from wealth that is in the hands of a tiny sector of the population. There were periods of American history, such as during the Gilded Age in the 1920s and the roaring 1990s, when something similar was going on. But the current period is extreme because inequality comes from super-wealth. Literally, the top one-tenth of a percent are just super wealthy. This is not only extremely unjust in itself but represents a development that has corrosive effects on democracy and on the vision of a decent society.

What does all this mean in terms of the American Dream? Is it dead?

The "American Dream" was all about class mobility. You were born poor but could get out of poverty through hard work and provide a better future for your children. It was possible for some workers to find a decent-paying job, buy a home, a car, and pay for a kid's education. It's all collapsed—and we shouldn't have too many illusions about when it was partially real. Today social mobility in the United States is below other rich societies.

Is the United States then a democracy in name only?

The United States professes to be a democracy, but it has clearly become something of a plutocracy, although it is still an open and free society by comparative standards. But let's be clear about what democracy means. In a democracy, the public influences policy and then the government carries out actions determined by the public. For the most part, the US government carries out actions that benefit corporate and financial interests. It is also important to understand that privileged and powerful sectors in society have never liked democracy, for good reasons. Democracy places power in the hands of the population and takes it away from them. In fact, the privileged and powerful classes of this country have always sought to find ways to limit power from being placed in the hands of the general population—and they are breaking no new ground in this regard.

Concentration of wealth yields to concentration of power. I think this is an undeniable fact. And since capitalism always leads in the end to concentration of wealth, doesn't it follow that capitalism is antithetical to democracy?

Concentration of wealth leads naturally to concentration of power, which in turn translates to legislation favoring the interests of the rich and powerful and thereby increasing even further the concentration of power and wealth. Various political measures, such as fiscal policy, deregulation, and rules for corporate governance, are designed to increase the concentration of wealth and power. And that's what we've been seeing during the neoliberal era. It is a vicious cycle in constant progress. The state is there to provide security and support to the interests of the privileged and powerful sectors in society, while the rest of the population is left to experience the brutal reality of capitalism. Socialism for the rich, capitalism for the poor.

So, yes, in that sense capitalism actually works to undermine democracy. But what has just been described—that is, the vicious cycle of concentration of power and wealth—is so traditional that it is even described by Adam Smith in 1776. He says in his famous *Wealth of Nations* that, in England, the people who own society, in his days the merchants and the manufacturers, are "the principal architects of policy." And they make sure that their interests are very well cared for, however grievous the impact of the policies they advocate and implement through government is on the people of England or others.

Now, it's not merchants and manufacturers who own society and dictate policy. It is financial institutions and multinational corporations. Today they are the groups that Adam Smith called *the masters of mankind.* And they are following the same vile maxim that he formulated: *All for ourselves and nothing for anyone else.* They will pursue policies that benefit them and harm everyone else because capitalist interests dictate that they do so. It's in the nature of the system. And in the absence of a general, popular reaction, that's pretty much all you will get.

Let's return to the idea of the American Dream and talk about the origins of the American political system. I mean, it was never intended to be a democracy (actually the term always used to describe the architecture

of the American political system was "republic," which is very different from a democracy, as the ancient Romans well understood), and there had always been a struggle for freedom and democracy from below, which continues to this day. In this context, wasn't the American Dream built at least partly on a myth?

Sure. Right through American history, there's been an ongoing clash between pressure for more freedom and democracy coming from below and efforts at elite control and domination from above. It goes back to the founding of the country, as you pointed out. The "founding fathers," even James Madison, the main framer, who was as much a believer in democracy as any other leading political figure in those days, felt that the United States' political system should be in the hands of the wealthy because the wealthy are the "more responsible set of men." And, thus, the structure of the formal constitutional system placed more power in the hands of the Senate, which was not elected in those days. It was selected from the wealthy men who, as Madison put it, had sympathy for the owners of wealth and private property.

This is clear when you read the debates of the Constitutional Convention. As Madison said, a major concern of the political order has to be "to protect the minority of the opulent against the majority." And he had arguments. If everyone had a vote freely, he said, the majority of the poor would get together and they would organize to take away the property of the rich. That, he added, would be obviously unjust, so the constitutional system had to be set up to prevent democracy.

Recall that Aristotle had said something similar in his *Politics*. Of all political systems, he felt that democracy was the best. But he saw the same problem that Madison saw in a true democracy, which is that the poor might organize to take away the property of the rich. The solution that he proposed, however, was something like a welfare state with the aim of reducing economic inequality. The other alternative, pursued by the "founding fathers," is to reduce democracy.

Now, the so-called American Dream was always based partly in myth and partly in reality. From the early nineteenth century onward and up until fairly recently, working-class people, including immigrants, had expectations that their lives would improve in American society through hard work. And that was partly true, although it did not apply for the most part to African Americans and women until much later. This no longer seems to be the case. Stagnating incomes, declining living standards, outrageous student debt levels, and hard-to-come-by decent-paying jobs have created a sense of hopelessness among many Americans, who are beginning to look with certain nostalgia toward the past. This explains, to a great extent, the rise of the likes of Donald Trump and the appeal among the youth of the political message of someone like Bernie Sanders.

After World War II, and pretty much up until the mid-1970s, there was a movement in the United States in the direction of a more egalitarian society and toward greater freedom, in spite of great resistance and oppression from the elite and various government agencies. What happened afterward that rolled back the economic progress of the postwar era, creating in the process a new socioeconomic order that has come to be identified as that of neoliberalism?

Beginning in the 1970s, partly because of the economic crisis that erupted in the early years of that decade and the decline in the rate of profit, but also partly because of the view that democracy had become too widespread, an enormous, concentrated, coordinated business offensive was begun to try to beat back the egalitarian efforts of the postwar era, which only intensified as time went on. The economy itself shifted to financialization. Financial institutions expanded enormously. By 2007, right before the crash for which they had considerable responsibility, financial institutions accounted for a stunning 40 percent of corporate profit. A vicious cycle between concentrated capital and politics accelerated, while increasingly wealth concentrated in the financial sector. Politicians, faced with the rising cost of campaigns, were driven ever deeper into the pockets of wealthy backers. And politicians rewarded them by pushing policies favorable to Wall Street and other powerful business interests. Throughout this period, we have a renewed form of class warfare directed by the business class against the working people and the poor, along with a conscious attempt to roll back the gains of the previous decades.

Now that Trump is the president-elect, is the Bernie Sanders political revolution over?

That's up to us and others to determine. The Sanders "political revolution" was quite a remarkable phenomenon. I was certainly surprised, and pleased. But we should remember that the term "revolution" is somewhat misleading. Sanders is an honest and committed New Dealer. The fact that he's considered "radical" tells us how far the elite political spectrum has shifted to the right during the neoliberal period. There have been some promising offshoots of the Sanders mobilization, like the Brand New Congress movement and several others.

There could, and should, also be efforts to develop a genuine independent left party, one that doesn't just show up every four years but is working constantly at the grassroots, both at the electoral level (everything from school boards to town meetings to state legislatures and on up) and in all the other ways that can be pursued. There are plenty of opportunities—and the stakes are substantial, particularly when we turn attention to the two enormous shadows that hover over everything: nuclear war and environmental catastrophe, both ominous, demanding urgent action.

The US Health System Is an International Scandal—and ACA Repeal Will Make It Worse

C. J. POLYCHRONIOU: Trump and the Republicans are bent on doing away with Obamacare. Doesn't the 2010 Patient Protection and Affordable Care Act (ACA) represent an improvement over what existed before? And, what would the Republicans replace it with?

NOAM CHOMSKY: I perhaps should say, to begin, that I have always felt a little uncomfortable about the term "Obamacare." Did anyone call Medicare "Johnsoncare?" Maybe wrongly, but it has seemed to me to have a tinge of Republican-style vulgar disparagement, maybe even of racism. But put that aside. Yes, the ACA is a definite improvement over what came before—which is not a great compliment. The US health care system has long been an international scandal, with about twice the per capita expenses of other wealthy (OECD) countries and relatively poor outcomes. The ACA did, however, bring improvements, including insurance for tens of millions of people who lacked it, banning of refusal of insurance for people with prior disabilities, and other gains—and also, it appears to have led to a reduction in the increase of health care costs, though that is hard to determine precisely.

Originally published in *Truthout*, January 12, 2017

The House of Representatives, dominated by Republicans (with a minority of voters), has voted over fifty times in the past six years to repeal or weaken Obamacare, but they have yet to come up with anything like a coherent alternative. That is not too surprising. Since Obama's election, the Republicans have been pretty much the party of NO. Chances are that they will now adopt a cynical, Paul Ryan–style evasion, repeal and delay, to pretend to be honoring their fervent pledges while avoiding at least for a time the consequences of a possible major collapse of the health system and ballooning costs. It's far from certain. It's conceivable that they might patch together some kind of plan, or that the ultra-right and quite passionate "Freedom Caucus" may insist on instant repeal without a plan, damn the consequence for the budget, or, of course, for people.

One part of the health system that is likely to suffer is Medicaid, probably through block grants to states, which gives the Republican-run states opportunities to gut it. Medicaid only helps poor people who "don't matter" and don't vote Republican anyway. So, according to Republican logic, why should the rich pay taxes to maintain it?

Article 25 of the UN Universal Declaration on Human Rights (UDHR) states that the right to health care is indeed a human right. Yet, it is estimated that close to 30 million Americans remain uninsured even with the ACA in place. What are some of the key cultural, economic, and political factors that make the United States an outlier in the provision of free health care?

First, it is important to remember that the United States does not accept the Universal Declaration of Human Rights—though in fact the UDHR was largely the initiative of Eleanor Roosevelt, who chaired the commission that drafted its articles, with quite broad international participation.

The UDHR has three components, which are of equal status: civil-political, socioeconomic, and cultural rights. The United States formally accepts the first of the three, though it has often violated its provisions. The United States pretty much disregards the third. And to the point here, the United States has officially and strongly condemned the second component, socioeconomic rights, including Article 25.

Opposition to Article 25 was particularly vehement in the Reagan and

Bush I years. Paula Dobriansky, deputy assistant secretary of state for human rights and humanitarian affairs in these administrations, dismissed the "myth" that "economic and social rights constitute human rights," as the UDHR declares. She was following the lead of Reagan's UN ambassador Jeane Kirkpatrick, who ridiculed the myth as "little more than an empty vessel into which vague hopes and inchoate expectations can be poured." Kirkpatrick thus joined Soviet Ambassador Andrei Vyshinsky, who agreed that it was a mere "collection of pious phrases." The concepts of Article 25 are "preposterous" and even a "dangerous incitement," according to ambassador Morris Abram, the distinguished civil rights attorney who was US Representative to the UN Commission on Human Rights under Bush I, casting the sole veto of the UN Right to Development, which closely paraphrased Article 25 of the UDHR. The Bush II administration maintained the tradition by voting alone to reject a UN resolution on the right to food and the right to the highest attainable standard of physical and mental health (the resolution passed 52–1).

Rejection of Article 25, then, is a matter of principle. And also a matter of practice. In the OECD ranking of social justice, the United States is in twenty-seventh place out of thirty-one, right above Greece, Chile, Mexico, and Turkey.[1] This is happening in the richest country in world history, with incomparable advantages. It was quite possibly already the richest region in the world in the eighteenth century.

In extenuation of the Reagan-Bush-Vyshinsky alliance on this matter, we should recognize that formal support for the UDHR is all too often divorced from practice.

US dismissal of the UDHR in principle and practice extends to other areas. Take labor rights. The United States has failed to ratify the first principle of the International Labour Organization Convention, which endorses "Freedom of Association and Protection of the Right to Organise." An editorial comment in the *American Journal of International Law* refers to this provision of the International Labour Organization Convention as "the untouchable treaty in American politics." US rejection is guarded with such fervor, the report continues, that there has never even been any debate about the matter. The rejection of the International Labour Organization Convention contrasts dramatically with the fervor of Washington's dedication to the highly protectionist elements of the misnamed "free trade agreements," designed to guarantee monopoly pricing rights for corporations ("intellectual property rights"), on spurious grounds. In general, it would be more accurate to call these "investor rights agreements."

Comparison of the attitude toward elementary rights of labor and extraordinary rights of private power tells us a good deal about the nature of American society.

Furthermore, US labor history is unusually violent. Hundreds of US workers were being killed by private and state security forces in strike actions, practices unknown in similar countries. In her history of American labor, Patricia Sexton—noting that there are no serious studies—reports an estimate of seven hundred strikers killed and thousands injured from 1877 to 1968, a figure which, she concludes, may "grossly understate the total casualties." In comparison, one British striker was killed since 1911.

As struggles for freedom gained victories and violent means became less available, business turned to softer measures, such as the "scientific methods of strike breaking" that have become a leading industry. In much the same way, the overthrow of reformist governments by violence, once routine, has been displaced by "soft coups" such as the recent coup in Brazil, though the former options are still pursued when possible, as in Obama's support for the Honduran military coup in 2009, in near isolation. Labor remains relatively weak in the United States in comparison to similar societies. It is constantly battling even for survival as a significant organized force in the society, under particularly harsh attack since the Reagan years.

All of this is part of the background for the US departure in health care from the norm of the OECD, and even less privileged societies. But there are deeper reasons why the United States is an "outlier" in health care and social justice generally. These trace back to unusual features of American history. Unlike other developed state capitalist industrial democracies, the political economy and social structure of the United States developed in a kind of tabula rasa. The expulsion or mass killing of Indigenous nations cleared the ground for the invading settlers, who had enormous resources and ample fertile lands at their disposal, and extraordinary security for reasons of geography and power. That led to the rise of a society of individual farmers, and also, thanks to slavery, substantial control of the product that fueled the industrial revolution: cotton, the foundation of manufacturing, banking, commerce, retail for both the United States and Britain and, less directly, for other European societies. Also relevant is the fact that the country has actually been at war for five hundred years with little respite, a history that has created "the richest, most powerful, and ultimately most militarized nation in world history," as scholar Walter Hixson has documented.[2]

For similar reasons, American society lacked the traditional social stratification and autocratic political structure of Europe, and the various measures of social support that developed unevenly and erratically. There has been ample state intervention in the economy from the outset—dramatically in recent years—but without general support systems.

As a result, US society is, to an unusual extent, business-run, with a highly class-conscious business community dedicated to "the everlasting battle for the minds of men." The business community is also set on containing or demolishing the "political power of the masses," which it deems as a serious "hazard to industrialists" (to sample some of the rhetoric of the business press during the New Deal years, when the threat to the overwhelming dominance of business power seemed real).

Here is yet another anomaly about US health care: according to data by the Organization for Economic Cooperation and Development (OECD), the United States spends far more on health care than most other advanced nations, yet Americans have poor health outcomes and are plagued by chronic illnesses at higher rates than the citizens of other advanced nations. Why is that?

US health care costs are estimated to be about twice the OECD average, with rather poor outcomes by comparative standards. Infant mortality, for example, is higher in the United States than in Cuba, Greece, and the EU generally, according to CIA figures.

As for reasons, we can return to the more general question of social justice comparisons, but there are special reasons in the health care domain. To an unusual extent, the US health care system is privatized and unregulated. Insurance companies are in the business of making money, not providing health care, and when they undertake the latter, it is likely not to be in the best interests of patients or to be efficient. Administrative costs are far greater in the private component of the health care system than in Medicare, which itself suffers by having to work through the private system.

Comparisons with other countries reveal much more bureaucracy and higher administrative costs in the US privatized system than elsewhere. One study of the United States and Canada a decade ago, by medical researcher Steffie Woolhandler and associates, found enormous disparities and concluded: "Reducing U.S. administrative costs to Canadian levels would save at least $209 billion annually, enough to fund universal coverage." Another anomalous feature of the US system is the law banning the government from negotiating drug prices, which leads to highly inflated prices in the United States as compared with other countries. That effect is magnified considerably by the extreme patent rights accorded to the pharmaceutical industry in "trade agreements," enabling monopoly profits. In a profit-driven system, there are also incentives for expensive treatments rather than preventive care, as strikingly in Cuba, with remarkably efficient and effective health care.

Why aren't Americans demanding—not simply expressing a preference for in survey polls—access to a universal health care system?

They are indeed expressing a preference, over a long period. Just to give one telling illustration, in the late Reagan years 70 percent of the adult population thought that health care should be a constitutional guarantee, and 40 percent thought it already was in the Constitution since it is such an obviously legitimate right. Poll results depend on wording and nuance, but they have quite consistently, over the years, shown strong and often large majority support for universal health care—often called "Canadian-style," not because Canada necessarily has the best system, but because it is close by and observable. The early ACA proposals called for a "public option." It was supported by almost two-thirds of the population, but was dropped without serious consideration, presumably as part of a compact with financial institutions. The legislative bar to government negotiation of drug prices was opposed by 85 percent, also disregarded—again, presumably, to prevent opposition by the pharmaceutical giants. The preference for universal health care is particularly remarkable in light of the fact that there is almost no support or advocacy in sources that reach the general public and virtually no discussion in the public domain.

The facts about public support for universal health care receive occasional comment, in an interesting way. When running for president in 2004, Democrat John Kerry, the *New York Times* reported, "took pains . . . to say that his plan for expanding access to health insurance would not create a new government program," because "there is so little political support for government intervention in the health care market in the United States." At the same time, polls in the *Wall Street Journal*, *Businessweek*, the *Washington Post*, and other media found overwhelming public support for government guarantees to everyone of "the best and most advanced health care that technology can supply."

But that is only public support. The press reported correctly that there was little "political support" and that what the public wants is "politically impossible"—a polite way of saying that the financial and pharmaceutical industries will not tolerate it, and in American democracy, that's what counts.

Returning to your question, it raises a crucial question about American democracy: Why isn't the population "demanding" what it strongly prefers? Why is it allowing concentrated private capital to undermine necessities of life in the interests of profit and power? The "demands" are hardly utopian. They are commonly satisfied elsewhere, even in sectors of the US system. Furthermore, the demands could readily be implemented even without significant legislative breakthroughs. For example, by steadily reducing the age for entry to Medicare.

The question directs our attention to a profound democratic deficit in an atomized society, lacking the kind of popular associations and organizations that enable the public to participate in a meaningful way in determining the course of political, social, and economic affairs. These would crucially include a strong and participatory labor movement and actual political parties growing from public deliberation and participation instead of the elite-run candidate-producing groups that pass for political parties. What remains is a depoliticized society in which a majority of voters (barely half the population even in the super-hyped presidential elections, much less in others) are literally disenfranchised, in that their representatives disregard their preferences while effective decision-making lies largely in the hands of tiny concentrations of wealth and corporate power, as study after study reveals.

The prevailing situation reminds us of the words of America's leading twentieth-century social philosopher, John Dewey, much of whose work focused on democracy and its failures and promise. Dewey deplored the domination by "business for private profit through private control of banking, land, industry, reinforced by command of the press, press agents and other means of publicity and propaganda" and recognized that "power today resides in control of the means of production, exchange, publicity, transportation and communication. Whoever owns them rules the life of the country," even if democratic forms remain. Until those institutions are in the hands of the public, he continued, politics will remain "the shadow cast on society by big business."

This was not a voice from the marginalized far left, but from the mainstream of liberal thought.

Turning finally to your question again, a rather general answer, which applies in its specific way to contemporary western democracies, was provided by David Hume over 250 years ago, in his classic study *Of the First Principles of Government*. Hume found

> nothing more surprising than to see the easiness with which the many are governed by the few; and to observe the implicit submission with which men resign their own sentiments and passions to those of their rulers. When we enquire by what means this wonder is brought about, we shall find, that as Force is always on the side of the governed, the governors have nothing to support them but opinion. 'Tis therefore, on opinion only that government is founded; and this maxim extends to the most despotic and most military governments, as well as to the most free and most popular.

Implicit submission is not imposed by laws of nature or political theory. It is a choice, at least in societies such as ours, which enjoys the legacy provided by the struggles of those who came before us. Here power is indeed "on the side of the governed," if they organize and act to gain and exercise it. That holds for health care and for much else.

The Perils of Market-Driven Education*

C. J. POLYCHRONIOU: At least since the Enlightenment, education has been seen as one of the few opportunities for humanity to lift the veil of ignorance and create a better world. What are the actual connections between democracy and education, or are those links based mainly on a myth, as Neil Postman argued in *The End of Education*?

NOAM CHOMSKY: I don't think there is a simple answer. The actual state of education has both positive and negative elements, in this regard. An educated public is surely a prerequisite for a functioning democracy—where "educated" means not just informed but enabled to inquire freely and productively, the primary end of education. That goal is sometimes advanced, sometimes impeded, in actual practice, and to shift the balance in the right direction is a major task—a task of unusual importance in the United States, in part because of its unique power, in part because of ways in which it differs from other developed societies.

It is important to remember that although the richest country in the world for a long time, until World War II, the United States was something of a cultural backwater. If one wanted to study advanced science or math, or to become a writer and artist, one would often be attracted to Europe. That changed with World War II for obvious reasons, but only for part of the population. To take what is arguably the most important question in

---

* Coauthored with Lily Sage; originally published in *Truthout*, October 22, 2016

human history, how to deal with climate change, one impediment is that in the United States, 40 percent of the population sees it as no problem because Christ will return within the next few decades—symptomatic of many other pre-modern features of the society and culture.

Much of what prevails in today's world is market-driven education, which is actually destroying public values and undermining the culture of democracy with its emphasis on competition, privatization, and profit-making. As such, what model of education do you think holds the best promise for a better and peaceful world?

In the early days of the modern educational system, two models were sometimes counterposed. Education could be conceived as a vessel into which one pours water—and a very leaky vessel, as we all know. Or it could be thought of as a thread, laid out by the instructor, along which students proceed in their own ways, developing their capacities to "inquire and create"—the model advocated by Wilhelm von Humboldt, the founder of the modern university system.

The educational philosophies of John Dewey, Paulo Freire, and other advocates of progressive and critical pedagogy can, I think, be regarded as further developments of the Humboldtian conception—which is often implemented as a matter of course in research universities, because it is so essential to advanced teaching and research, particularly in the sciences. A famous MIT physicist was known for telling his freshman courses that it doesn't matter what we cover, it matters what you discover.

The same ideas have been quite imaginatively developed down to the kindergarten level, and they are quite appropriate everywhere in the educational system, and of course not just in the sciences. I was personally lucky to have been in an experimental Deweyite school until I was twelve, a highly rewarding experience, quite different from the academic high school I attended, which tended toward the water-in-a-vessel model, as do currently fashionable programs of teach-to-test. The alternative ones are the kinds of models that should be pursued if there is to be some hope that a truly educated population, in all of the dimensions of the term, can face the very critical questions that are right now on the agenda.

The market-driven education tendencies that you mention are unfortunately very real, and harmful. They should, I think, be regarded as part of the general neoliberal assault on the public. The business model seeks "efficiency," which means imposing "flexibility of labor" and what Alan Greenspan hailed as "growing worker insecurity" when he was praising the great economy he was running (before it crashed). That translates into such measures as undermining longer-term commitments to faculty and relying on cheap and easily exploitable temporary labor (adjuncts, graduate students). The consequences are harmful to the work force, the students, research and inquiry, in fact all the goals that higher education should seek to achieve.

Sometimes such attempts to drive the higher education system toward service to the private sector take forms that are almost comical. In the state of Wisconsin, for example, governor Scott Walker and other reactionaries have been attempting to undermine what was once the great University of Wisconsin, changing it to an institution that will serve the needs of the business community of Wisconsin, while also cutting the budget and increasing reliance on temporary staff ("flexibility"). At one point the state government even wanted to change the traditional mission of the university, deleting the commitment to "seeking truth"—a waste of time for an institution producing people who will be useful for Wisconsin businesses. That was so outrageous that it hit the newspapers, and they had to claim it was a clerical error and withdraw it.

It is, however, illustrative of what is happening, not only in the United States but also in many other places. Commenting on these developments in the UK, Stefan Collini concluded all too plausibly that the Tory government is attempting to turn first-class universities into third-class commercial institutions. So, for example, the classics department at Oxford will have to prove that it can sell itself on the market. If there is no market demand, why should people study and investigate classical Greek literature? That's the ultimate vulgarization that can result from imposing the state capitalist principles of the business classes on the whole of society.

What needs to be done in order to provide a system of free higher education in the United States and, by extension, divert funding from the military-industrial complex and the prison-industrial complex into education? Would this require a national identity crisis on the part of a

historically expansionist, interventionist, and racist nation?

I don't feel that the issue runs that deep. The United States was no less expansionist, interventionist, racist in earlier years, but it nevertheless was in the forefront of developing mass public education. And though the motives were sometimes cynical—turning independent farmers into cogs in mass production industry, something they bitterly resented—nevertheless there were many positive aspects to these developments. In more recent years, higher education was virtually free. After World War II, the GI Bill provided tuition and even subsidies to millions of people who would probably never have gone to college, which was highly beneficial to them and contributed to the great postwar growth period. Even private colleges had very low fees by contemporary standards. And the country then was far poorer than it is today. Elsewhere higher education is free or close to it in rich countries like Germany (the most respected country in the world, according to polls) and Finland (which consistently ranks high in achievement) and much poorer countries like Mexico, which has a high-quality higher education system. Free higher education could be instituted without major economic or cultural difficulties, it seems. The same is true of a rational public health system like that of comparable countries.

During the industrial era, many working-class people throughout the capitalist world immersed themselves in the study of politics, history, and political economy through a process of informal education as part of their effort to understand and change the world through the class struggle. Today, the situation looks vastly different, with much of the working-class population having embraced empty consumerism and political indifference, or, worse, often enough supporting political parties and candidates who are in fact staunch supporters of corporate and financial capitalism and advance an anti–working class agenda. How do we explain this radical shift in working-class consciousness?

The change is as clear as it is unfortunate. Quite commonly these efforts were based in unions and other working-class organizations, with participation of intellectuals in left parties—all victims of Cold War repression and propaganda and the bitter class conflict waged by the business classes against labor and popular organization, mounting particularly during the neoliberal period.

It is worth remembering the early years of the Industrial Revolution. The working-class culture of the time was alive and flourishing. There's a great book about the topic by Jonathan Rose, called *The Intellectual Life of the British Working Class*. It's a monumental study of the reading habits of the working class of the day. He contrasts "the passionate pursuit of knowledge by proletarian autodidacts" with the "pervasive philistinism of the British aristocracy." Pretty much the same was true in the new working-class towns in the United States, like eastern Massachusetts, where an Irish blacksmith might hire a young boy to read the classics to him while he was working. Factory girls were reading the best contemporary literature of the day, what we study as classics. They condemned the industrial system for depriving them of their freedom and culture. This went on for a long time.

I am old enough to remember the atmosphere of the 1930s. A large part of my family came from the unemployed working class. Many had barely gone to school. But they participated in the high culture of the day. They would discuss the latest plays, concerts of the Budapest String Quartet, different varieties of psychoanalysis, and every conceivable political movement. There was also a very lively workers' education system with which leading scientists and mathematicians were directly involved. A lot of this has been lost . . . but it can be recovered and it is not lost forever.

Part III

Anarchism, Communism, and Revolutions

C. J. POLYCHRONIOU: Noam, from the late nineteenth century to the mid- or even late twentieth century, anarchism and communism represented live and vital movements throughout the Western world but also in Latin America and certain parts of Asia and Africa. However, the political and ideological landscape seems to have shifted radically by the early to late 1980s to the point that, while resistance to capitalism remains ever present, it is largely localized and devoid of a vision about strategies for the founding of a new socioeconomic order. Why did anarchism and communism flourish at the time they did, and what are the key factors for their transformation from major ideologies to marginalized belief systems?

NOAM CHOMSKY: If we look more closely, I think we find that there are live and vital movements of radical democracy, often with elements of anarchist and communist ideas and participation, during periods of upheaval and turbulence, when—to paraphrase Gramsci—the old is tottering and the new is unborn but is offering tantalizing prospects. Thus, in late nineteenth-century America, when industrial capitalism was driving independent farmers and artisans to become an industrial proletariat, evoking plenty of bitter resistance, a powerful and militant labor movement arose dedicated to the principle that "those who work in the mills should own them" alongside a mass radical farmers' movement that sought to free farmers from the clutches of

Originally published in *Truthout*, July 17, 2016

banks and merchants. The dramatic era of decolonization also gave rise to radical movements of many kinds, and there are many other cases, including the 1960s. The neoliberal period since the '80s has been one of regression and marginalization for much of the world's population, but Karl Marx's old mole is never far from the surface and appears in unexpected places. The spread of worker-owned enterprises and cooperatives in the United States, while not literally anarchist or communist, carries seeds of far-reaching radical transformation, and it is not alone.

Anarchism and communism share close affinities but have also been mortal enemies since the time of Marx and Russian anarchist Mikhail Bakunin. Are their differences purely strategic about the transition from capitalism to socialism, or do they also reflect different perspectives about human nature and economic and social relations?

My feeling is that the picture is more nuanced. Thus left anti-Bolshevik Marxism often was quite close to anarchosyndicalism. Prominent left Marxists, like Karl Korsch, were quite sympathetic to the Spanish anarchist revolution. Daniel Guerin's book *Anarchism* verges on left Marxism. During his left period in mid-1917, Lenin's writings, notably *State and Revolution*, had a kind of anarchist tinge. There surely were conflicts over tactics and much more fundamental matters. Engels's critique of anarchism is a famous illustration. Marx had very little to say about postcapitalist society, but the basic thrust of his thinking about long-term goals seems quite compatible with major strains of anarchist thinking and practice.

Certain anarchist traditions, influenced by Bakunin, advocate violence as a means of bringing about social change, while others, influenced by Russian anarchist Peter Kropotkin, seem to regard violence not only politically ineffective in securing a just social order but morally indefensible. The communist tradition has also been divided over the use of violence even in situations where the conditions seem to have been ripe for revolutions. Can social revolutions take place without violence?

I don't see how there can be a general answer. Struggles to overcome class power and privilege are sure to be resisted, sometimes by force. Perhaps a point will come where violence in defense against forceful efforts to maintain power is warranted. Surely it is a last resort.

In your writings, you have maintained the view that the Soviet Union was never a socialist state. Do you accept the view that it was a "deformed workers state" or do you believe that it was a form of state capitalism?

The terms of political discourse are not models of precision. By the time the soviets and factory councils were eliminated—quite early on—there was hardly a trace of a "workers state." [Factory councils were forms of political and economic organization in which the place of work is controlled collectively by the workers.] The system had wage labor and other features of capitalism, so I suppose one could call it a kind of tyrannical state capitalism in some respects.

In certain communist circles, a distinction has been drawn between Leninism and Stalinism, while the more orthodox communists have argued that the Soviet Union began a gradual abandonment of socialism with the rise of Nikita Khrushchev to power. Can you comment on these two points of contention, with special emphasis in the alleged differences between Leninism and Stalinism?

I would place the abandonment of socialism much earlier, under Lenin and Trotsky, at least if socialism is understood to mean at a minimum control by working people over production. The seeds of Stalinism were present in the early Bolshevik years, partly attributable to the exigencies of the civil war and foreign invasion, partly to Leninist ideology. Under Stalin it became a monstrosity.

Faced with the challenges and threats (both internal and external) that it did face following the takeover of power, did the Bolsheviks have any other option than centralizing power, creating an army, and defending the October Revolution by any means necessary?

It is more appropriate, I think, to ask whether the Bolsheviks had any other option for defending their power. By adopting the means they chose, they destroyed the achievements of the popular revolution. Were there alternatives?

I think so, but the question takes us into difficult and contested territory. It's possible, for example, that instead of ignoring Marx's ideas in his later years about the revolutionary potential of the Russian peasantry, they might have pursued them and offered support for peasant organizing and activism instead of marginalizing it (or worse). And they could have energized rather than undermined the soviets and factory councils. But all that raises many questions, both of fact and of speculation about possibilities—for example, about creating a disciplined and effective Red Army, choice of guerrilla versus conventional military tactics, political versus military warfare, and much else.

Would you accept the view that the labor concentration camps and the other horrible crimes that took place under Stalin's reign are unlikely to have taken place if either Lenin or Trotsky were in power instead?

I strongly doubt that Lenin or Trotsky would have carried out crimes anything like these.

And how do you see the Maoist revolution? Was China at any point a socialist state?

The "Maoist revolution" was a complex affair. There was a strong popular element in early Chinese Marxism, discussed in illuminating work by Maurice Meisner. William Hinton's remarkable study *Fanshen* captures vividly a moment of profound revolutionary change, not just in social practices but in the mentality and consciousness of the peasants, with party cadres often submitting to popular control, according to his account. Later the totalitarian system was responsible for horrendous crimes, notably the "Great Leap Forward," with its huge death toll in the tens of millions. Despite these crimes, as economists Amartya Sen and Jean Dreze demonstrate, from independence until 1979, when the Deng reforms began, Chinese programs of rural health and development saved the lives of 100 million people in comparison to India in the same years. What any of this has to do with socialism depends on how one interprets that battered term.

Cuba under Castro?

In assessing developments in Cuba since it achieved independence under

Castro in January 1959, one cannot overlook the fact that from almost the first moment, Cuba was subjected to vicious attack by the global superpower. By late 1959, planes based in Florida were bombing Cuba. By March, a secret decision was made to overthrow the government. The incoming Kennedy administration carried out the Bay of Pigs invasion. Its failure led to near hysteria in Washington, and Kennedy launched a war to bring "the terrors of the earth" to Cuba, in the words of his close associate, historian Arthur Schlesinger, in his semiofficial biography of Robert Kennedy, who was placed in charge of the operation as his highest priority. It was no small affair, and was one of the factors that led to the missile crisis, which Schlesinger rightly described as the most dangerous moment in history. After the crisis, the terrorist war resumed. Meanwhile, a crushing embargo was imposed, which took a huge toll on Cuba. It continues to this day, opposed by virtually the entire world.

When Russian aid ended, Clinton made the embargo harsher, and a few years later, the Helms-Burton Act made it harsher still. The effects have of course been very severe. They are reviewed in a comprehensive study by Salim Lamrani. Particularly onerous has been the impact on the health system, deprived of essential medical supplies. Despite the attack, Cuba has developed a remarkable health system, and has an unmatched record of medical internationalism—as well as playing a crucial role in the liberation of Black Africa and ending the apartheid regime in South Africa. There have also been severe human rights violations, though nothing like what has been standard in the US-dominated countries of the region or the US-backed national security states of South America. And, of course, the worst human rights violations in Cuba in recent years have been in Guantanamo, which the United States took from Cuba at gunpoint in the early twentieth century and refuses to return. Overall, a mixed story, and not easy to evaluate, given the complex circumstances.

Overall, do you regard the collapse of so-called actually existing socialism as a positive outcome, and, if so, why? In what ways has this development been beneficial to the socialist vision?

When the Soviet Union collapsed I wrote an article describing the events as a small victory for socialism, not only because of the fall of one of the most antisocialist states in the world, where working people had fewer rights than in the West, but also because it freed the term "socialism" from the burden of being associated in the propaganda systems of East and West with Soviet tyranny—for the East, in order to benefit from the aura of authentic socialism, for the West, in order to demonize the concept.

My argument on what came to be known as "actually existing socialism" has been that the Soviet state attempted since its origins to harness the energies of its own population and oppressed people elsewhere in the service of the men who took advantage of the popular ferment in Russia in 1917 to seize state power.

Since its origins, socialism has meant the liberation of working people from exploitation. As the Marxist theoretician Anton Pannekoek observed, "This goal is not reached and cannot be reached by a new directing and governing class substituting itself for the bourgeoisie," but can only be "realized by the workers themselves being master over production." Mastery over production by the producers is the essence of socialism, and means to achieve this end have regularly been devised in periods of revolutionary struggle, against the bitter opposition of the traditional ruling classes and the "revolutionary intellectuals" guided by the common principles of Leninism and Western managerialism, as adapted to changing circumstances. But the essential element of the socialist ideal remains: to convert the means of production into the property of freely associated producers and thus the social property of people who have liberated themselves from exploitation by their master, as a fundamental step toward a broader realm of human freedom.

The Leninist intelligentsia had a different agenda. They fit Marx's description of the "conspirators" who "preempt the developing revolutionary process" and distort it to their ends of domination. "Hence their deepest disdain for the more theoretical enlightenment of the workers about their class interests," which included the overthrow of the Red Bureaucracy of which Bakunin warned, and the creation of mechanisms of democratic control over production and social life. For the Leninist, the masses must be strictly disciplined, while the socialist will struggle to achieve a social order in which discipline "will become superfluous" as the freely associated producers "work for their own accord" (Marx). Libertarian socialism, furthermore, does not limit its aims to democratic control by producers over production, but seeks to abolish all forms of domination and hierarchy in every aspect of social and personal life—an unending struggle, since progress in achieving a more just society will lead to new insight and understanding of forms of oppression that may be concealed in traditional practice and consciousness.

The Leninist antagonism to the most essential features of socialism was evident from the very start. In revolutionary Russia, soviets and factory committees developed as instruments of struggle and liberation, with many flaws but with a rich potential. Lenin and Trotsky, upon assuming power, immediately devoted themselves to destroying the liberatory potential of these instruments, establishing the rule of the Communist Party—in practice, its Central Committee and its Maximal Leaders—exactly as Trotsky had predicted years earlier, as Rosa Luxemburg and other left Marxists warned at the time, and as the anarchists had always understood. Not only the masses but even the party must be subject to "vigilant control from above," so Trotsky held as he made the transition from revolutionary intellectual to state priest. Before seizing state power, the Bolshevik leadership adopted much of the rhetoric of people who were engaged in the revolutionary struggle from below, but their true commitments were quite different. This was evident before and became crystal clear as they assumed state power in October 1917.

A historian sympathetic to the Bolsheviks, E. H. Carr, writes that "the spontaneous inclination of the workers to organize factory committees and to intervene in the management of the factories was inevitably encouraged by a revolution which *led the workers to believe* that the productive machinery of the country belonged to them and could be operated by them at their own discretion and to their own advantage" [my emphasis]. For the workers, as one anarchist delegate said, "The factory committees were cells of the future. . . . They, not the state, should now administer."

But the state priests knew better, and moved at once to destroy the factory committees and to reduce the soviets to organs of their rule. On November 3, Lenin announced in a "Draft Decree on Workers' Control" that delegates elected to exercise such control were to be "answerable to the state for the maintenance of the strictest order and discipline and for the protection of property." As the year ended, Lenin noted that "we passed from workers' control to the creation of the Supreme Council of National Economy," which was to "replace, absorb and supersede the machinery of workers' control" (Carr). "The very idea of socialism is embodied in the concept of workers' control," one Menshevik trade unionist lamented. The Bolshevik leadership expressed the same lament in action, by demolishing the very idea of socialism.

Is the United States Ready for Socialism?

C. J. POLYCHRONIOU: Noam, the rise of the likes of Donald Trump and Bernie Sanders seems to indicate that US society is at the present moment in the midst of a major ideological readjustment brought about by the deteriorating state of the standard of living, the explosive growth of income inequality, and myriad other economic and social ills facing the country in the New Gilded Era. In your view, and given the peculiarities of US political culture, how significant are the 2016 presidential elections?

NOAM CHOMSKY: The elections are quite significant, whatever the outcome, in revealing the growing discontent and anger about the impact of the neoliberal programs of the past generation, which, as elsewhere quite generally, have had a harsh impact on the mass of the population while undermining functioning democracy and enriching and empowering a tiny minority, largely in financial industries that have a dubious, if not harmful, role in the economy. Similar developments are taking place, for similar reasons, in Europe. The tendencies have been clear for some time, but, in this election, the party establishments have lost control for the first time.

On the Republican side, in previous primaries they were able to eliminate candidates that arose from the base and to nominate their own man. But not this time, and they are desperate about the failure. On the Democratic side, the Sanders challenge and its success are no less unanticipated

Originally published in *Truthout*, May 18, 2016

than the Trump triumph and reflect similar disillusionment and concerns, very differently expressed but with some common elements. Trump supporters include much of the white working class. One can understand their anger and frustration, and why Trump's rhetoric might appeal to them. But they are betting on the wrong horse. His policy proposals—to the limited extent that they are coherent—not only do not seriously address their legitimate concerns but would be quite harmful to them. And not just to them.

Following somewhat on the footsteps of the Occupy Wall Street movement, Bernie Sanders has made economic inequality and social rights themes of his campaign. Is this trend likely to continue after the election, or will the momentum for reform fade away?

That's up to us, and, specifically, up to those who have been mobilized by the campaign, and to Sanders himself. The energy and commitment could fade away, like the Rainbow Coalition. Or it could become a continuing and growing force that is not focused on electoral extravaganzas even though it may use them to carry its concerns forward. That will be a critical choice in the coming months.

Is Bernie Sanders merely a New Dealer, or perhaps a European social democrat, or something further to the left?

He seems to me a decent and honest New Dealer—which is not so different from European social democracy (actually, both terms cover a pretty broad range).

In your view, are Keynesianism and social democracy still relevant and applicable in today's global economic environment, or simply defunct?

I think they are quite relevant, to restore some degree of sanity and decency to social and economic life—but not sufficient. We should aim well beyond.

Should the left in the United States fight for reforms along the lines of those articulated by Bernie Sanders, or should it devote itself to promoting a more radical version of social and economic change?

I don't think this has to be a choice, though of course the degree of emphasis on one or the other is a choice. Both can be pursued simultaneously, and can be mutually reinforcing. Take a venerable anarchist journal like *Freedom*, founded by Russian activist and philosopher Peter Kropotkin. Its pages are often devoted to ongoing social struggles with reformist aims, which would improve people's lives and create the basis for moving on. These concerns are guided by far more radical long-term objectives.

While supporting valuable reforms and efforts to protect and extend rights, there is no reason not to follow Russian anarchist Mikhail Bakunin's advice to create the germs of a future society within the present one, at the very same time. For example, we can support health and safety standards in the capitalist workplace while at the same time establishing enterprises owned and managed by the workforce. And even support for the reformist measures can (and should be) designed so as to highlight the roots of the problems in the existing institutions, encouraging the recognition that defending and expanding rights is just a step toward eliminating those roots.

Historically, one of the major challenges facing the labor movement in the United States is the absence of a national class-based political organization. Do you see this changing any time soon on account of the ideas of socialism beginning to establish roots among certain segments of the American population, particularly among the youth?

US political history is rather unusual among the developed state capitalist societies. The political parties have not been class-based to the same extent as elsewhere. They have been regional in large part, a residue of the Civil War, which has still not ended. In the last election, for example, the red (Republican) states looked remarkably like the Confederacy—party names switched after the civil rights movement opened the way for Nixon's racist "Southern strategy." The parties have also been based on rather ad hoc coalitions, which blur any possible class lines further, leaving the two parties as basically factions of the ruling business party, in the familiar phrase.

There is no indication of that changing, and in the US system of "first past the post" and massive campaign expenditures, it is very hard to break the lock of the two political parties, which are not membership or participatory parties, but more candidate-producing and fundraising organizations, with somewhat different policy orientations (within a fairly narrow range). It is rather striking, for example, to see how easily the Democratic Party almost openly abandons the white working class, which drifts to the hands of their most bitter class enemy, the leadership and power base of the Republican Party.

On socialism establishing roots among the young, one has to be cautious. It's not clear that "socialism" in the current context means something different from New Deal–style welfare-state capitalism—which would, in fact, be a very healthy development in today's ugly context.

How should we define socialism in the twenty-first century?

Like other terms of political discourse, "socialism" is quite vague and broad in application. How we should define it depends on our values and goals. A good start, fitting well into the American context, would be the recommendations of America's leading twentieth-century social philosopher, John Dewey, who called for democratization of all aspects of political, economic, and social life. He held that workers should be "the masters of their own industrial fate," and that "the means of production, exchange, publicity, transportation and communication" should be under public control. Otherwise, politics will remain "the shadow cast on society by big business" and social policy will be geared to the interests of the masters. That's a good start. And is deeply rooted in significant strands of the society and its complex history.

A problem facing today's left is that, whenever it came to power, it capitulated in no time to capitalist forces and became immersed itself in the practices of corruption and the pursuit of power for the sake of power and material gains. We have seen it in Brazil, in Greece, in Venezuela, and elsewhere. How do you explain this?

That's been a very sad development. The causes vary, but the results are highly destructive. In Brazil, for example, the PT (Workers' Party) had enormous opportunities and could have been a force for transforming Brazil and leading the way for the whole continent, given Brazil's unique position. Though there were some achievements, the opportunities were squandered as the party leadership joined the rest of the elite in sinking into the abyss of corruption.

Although it was clear that Bernie Sanders could not win the Democratic nomination, he sought to stick around as a candidate until the convention. What was his aim in doing so?

The intention, I presume, was pretty much what he has been saying: to have a significant role in formulating the party platform at the convention. That doesn't seem to me to matter much; platforms are mostly rhetoric. What could be quite significant is something different: using the opportunity of the electoral enthusiasm, largely fostered by propaganda, to organize an ongoing and growing popular movement, not geared to the electoral cycle, which will be devoted to bringing about badly needed changes by direct action and other appropriate means.

If the American Dream is dead, as Donald Trump says it is, why do surveys continue to show that the majority of those interviewed say they still believe and even live the American Dream? Was the American Dream ever reality, or just a myth?

The "American Dream" was a very mixed story. It traces back to the nineteenth century, when free people could obtain land and pursue other opportunities in an expanding economy—thanks to annihilation of the Indigenous nations who populated the country and the huge contribution to the economy of the most vicious form of slavery that has yet existed.

In later years the "dream" took other forms, for some, and sometimes. Until European immigration was sharply cut in 1924 in order to block undesirables (mainly Italians and Jews), immigrants could hope to work their way into a rich society, with incomparable advantages. In the 1950s and 1960s, the great growth years of state capitalism, working people, including African Americans for a rare moment in the past half-millennium of bitter repression, could hope to get a decently paying union job with benefits, buy a house and a car, send their kids to college. That dream pretty much ended with the shift of the economy toward financialization and neoliberalism from the 1970s, accelerating under Reagan and since. But there is no reason to suppose that the traditional "dream," such as it was, is over, or that something much better, much more humane and just, is beyond our reach.

Why I Choose Optimism over Despair

C. J. POLYCHRONIOU: Noam, your book *What Kind of Creatures Are We?* (Columbia University Press, 2015) brings together your investigation into language and the mind and long-held views of yours on society and politics. Let me start by asking you as to whether you feel that the biolinguistic approach to language that you have developed in the course of the past fifty years or so is still open to further exploration and, if so, what sort of questions remain unanswered about the acquisition of language.

NOAM CHOMSKY: Not just me, by any means. Quite a few people. One of the real pioneers was the late Eric Lenneberg, a close friend from the early 1950s when these ideas were brewing. His book *Biological Foundations of Language* is an enduring classic.

The program is very much open to further exploration. There are unanswered questions right at the borders of inquiry, the kinds that are crucial for advancing what Tom Kuhn called "normal science." And questions that lie beyond are traditional and tantalizing.

One topic that is beginning to be open to serious investigation is the realization of the capacity for language and its use in the brain. That's very hard to study. Similar questions are extremely difficult even in the case of insects, and for humans, they are incomparably harder, not only because of the vastly greater complexity of the brain. We know a good deal about the

Originally published in *Truthout*, February 14, 2016

human visual system, but that is because it is much the same as the visual systems of cats and monkeys, and (rightly or not) we permit invasive experimentation with these animals. That is impossible for humans because the human language capacity is so isolated biologically. There are no relevant analogues elsewhere in the biological world—a fascinating topic in itself.

Nevertheless, new noninvasive technologies are beginning to provide important evidence, which sometimes even is beginning to bear on open questions about the nature of language in interesting ways. These are among the topics at the borders of inquiry, along with a huge and challenging mass of problems about the properties of language and the principles that explain them. Lying far, far beyond—maybe even beyond human reach—are the kinds of questions that animated traditional thought (and wonder) about the nature of language, including such great figures as Galileo, Descartes, von Humboldt, and others: primary among them, what has been called "the creative aspect of language use," the ability of every human to construct in the mind and comprehend an unbounded number of new expressions expressing their thoughts, and to use them in ways appropriate to but not caused by circumstances, a crucial distinction.

We are "incited and inclined" but not "compelled," in Cartesian terminology. These are not matters restricted to language, by any means. The issue is put graphically by two leading neuroscientists who study voluntary motion, Emilio Bizzi and Robert Ajemian. Reviewing the current state of the art, they observe that we are beginning to understand something about the puppet and the strings, but the puppeteer remains a total mystery. Because of its centrality to our lives, and its critical role in constructing, expressing, and interpreting thought, the normal use of language illustrates these mysterious capacities in a particularly dramatic and compelling way. That is why normal language use, for Descartes, was a primary distinction between humans and any animal or machine, and a basis for his mind-body dualism—which, contrary to what is often believed, was a legitimate and sensible scientific hypothesis in his day, with an interesting fate.

What would you say is the philosophical relevance of language?

The comments above begin to deal with that question. It has been traditionally recognized that human language is a species property, common to humans apart from severe pathology, and unique to humans in essentials. One of Lenneberg's contributions was to begin to ground this radical discontinuity in sound modern biology, and the conclusion has only been strengthened by subsequent work (a matter that is hotly contested, but mistakenly so, I believe). Furthermore, work that Lenneberg also initiated reveals that the human language capacity appears to be dissociated quite sharply from other cognitive capacities. It is, furthermore, not only the vehicle of thought, but also probably the generative source of substantial parts of our thinking.

The close study of language also provides much insight into classical philosophical problems about the nature of concepts and their relation to mind-external entities, a matter much more intricate than often assumed. And more generally, it suggests ways to investigate the nature of human knowledge and judgment. In another domain, important recent work by John Mikhail and others has provided substantial support for some neglected ideas of John Rawls on relations of our intuitive moral theories to language structure. And much more. There is good reason why study of language has always been a central part of philosophical discourse and analysis, and new discoveries and insights, I think, bear directly on many of the traditional concerns.

The well-known University College London linguist Neil Smith argued in his book *Chomsky: Ideas and Ideals* (Cambridge University Press, 1999) that you put to rest the mind-body problematic not by showing that we have a limited understanding of the mind but that we cannot define what the body is. What can he possibly mean by this?

I wasn't the person who put it to rest. Far from it. Isaac Newton did. Early modern science, from Galileo and his contemporaries, was based on the principle that the world is a machine, a much more complex version of the remarkable automata then being constructed by skilled craftsmen, which excited the scientific imagination of the day, much as computers and information processing do today. The great scientists of the time, including Newton, accepted this "mechanical philosophy" (meaning the science of mechanics) as the foundation of their enterprise. Descartes believed he had pretty much established the mechanical philosophy, including all the phenomena of body, though he recognized that some phenomena lay beyond its reach, including, crucially, the "creative aspect of language use" described above. He therefore, plausibly, postulated a new principle—in the metaphysics of the day, a new substance, *res cogitans*, "thinking substance, mind." His followers devised experimental techniques to try to determine whether other creatures had this property, and, like Descartes, were concerned to discover how the two substances interacted.

Newton demolished the picture. He demonstrated that the Cartesian account of body was incorrect and, furthermore, that there could be no mechanical account of the physical world: the world is not a machine. Newton regarded this conclusion as so "absurd" that no one of sound scientific understanding could possibly entertain it—though it was true. Accordingly, Newton demolished the concept of body (material, physical, and so on), in the form that it was then understood, and there really is nothing to replace it, beyond "whatever we more or less understand." The Cartesian concept of mind remained unaffected. It has become conventional to say that we have rid ourselves of the mysticism of "the ghost in the machine." Quite the contrary: Newton exorcised the machine while leaving the ghost intact, a consequence understood very well by the great philosophers of the period, like John Locke.

Locke went on to speculate (in the accepted theological idiom) that just as God had added to matter properties of attraction and repulsion that are inconceivable to us (as demonstrated by "the judicious Mr. Newton"), so he might have "superadded" to matter the capacity of thought. The suggestion (known as "Locke's suggestion" in the history of philosophy) was pursued extensively in the eighteenth century, particularly by philosopher and chemist Joseph Priestley, adopted by Darwin, and rediscovered (apparently without awareness of the earlier origins) in contemporary neuroscience and philosophy.

There is much more to say about these matters, but that, in essence, is what Smith was referring to. Newton eliminated the mind-body problem in its classic Cartesian form (it is not clear that there is any other coherent version), by eliminating body, leaving mind intact. And in doing so, as David Hume concluded, "While Newton seemed to draw the veil from some of the mysteries of nature, he showed at the same time the imperfections of the mechanical philosophy . . . and thereby restored [nature's] ultimate secrets to that obscurity, in which they ever did and ever will remain."

When you made your breakthrough into the study of linguistics, B. F. Skinner's verbal behavior approach dominated the field and was widely employed in the field of marketing and promotions. Your critique of Skinner's approach not only overthrew the prevailing paradigm at the time but also established a new approach to linguistics. Yet, it seems that behavioralism still dominates the public realm when it comes to marketing and consumer behavior. Your explanation for this apparent antinomy?

Behavioral methods (though not exactly Skinner's) may work reasonably well in shaping and controlling thought and attitudes, hence some behavior, at least at the superficial level of marketing and inducing consumerism. The need to control thought is a leading doctrine of the huge PR industry, which developed in the freest countries in the world, Britain and the United States, motivated by the recognition that people had won too many rights to be controlled by force, so it was necessary to turn to other means: what one of the founders of the industry, Edward Bernays, called "the engineering of consent."

In his book *Propaganda*, a founding document of the industry, Bernays explained that engineering consent and "regimentation" were necessary in democratic societies so as to ensure that the "intelligent minority" will be able to act (of course, for the benefit of all) without the interference of the annoying public, who must be kept passive, obedient, and diverted; passionate consumerism is the obvious device, based on "creating wants" by various means.

As explained by his contemporary and fellow liberal intellectual Walter Lippmann, the leading public intellectual of the day, the "ignorant meddlesome outsiders"—the general public—must be "put in their place" as "spectators," not "participants," while "the responsible men" must be protected from "the trampling and the roar of a bewildered herd." This is an essential principle of prevailing democratic theory. Marketing to engineer consent by control of thought, attitudes, and behavior is a crucial lever to achieve these ends—and (incidentally) to keep profits flowing.

Many maintain the view that, as humans, we have a propensity for aggression and violence, which in actuality explains the rise of oppressive and repressive institutions that have defined much of human civilization throughout the world. How do you respond to this dark view of human nature?

Since oppression and repression exist, they are reflections of human nature. The same is true of sympathy, solidarity, kindness, and concern for others—and for some great figures, like Adam Smith, these were the essential properties of humans. The task for social policy is to design the ways we live and the institutional and cultural structure of our lives so as to favor the benign and to suppress the harsh and destructive aspects of our fundamental nature.

While it is true that humans are social beings and thus our behavior depends on the social and political arrangements in our lives, is there such a thing as a common good for all human beings that goes beyond basic aspirations like the need for food, shelter, and protection from external threats?

These are what Marx once called our "animal needs," which, he hoped, would be provided by realization of communism, freeing us to turn productively to our "human needs," which far transcend these in significance—though we cannot forget Brecht's admonition: "First, feed the face."

All in all, how would you define human nature—or, alternatively, what kind of creatures are we?

I open the book by saying that "I am not deluded enough to think I can provide a satisfactory answer" to this question—going on to say that "it seems reasonable to believe that in some domains at least, particularly with regard to our cognitive nature, there are insights of some interest and significance, some new, and that it should be possible to clear away some of the obstacles that hamper further inquiry, including some widely accepted doctrines with foundations that are much less stable than often assumed." I haven't become less deluded since.

You have defined your political philosophy as libertarian socialism/anarchism, but refuse to accept the view that anarchism as a vision of social order flows naturally from your views on language. Is the link then purely coincidental?

It's more than coincidental but much less than deductive. At a sufficient level of abstraction, there is a common element—which was sometimes recognized, or at least glimpsed, in the Enlightenment and Romantic eras. In both domains, we can perceive, or at least hope, that at the core of human nature is what Bakunin called "an instinct for freedom," which reveals itself both in the creative aspect of normal language use and in the recognition that no form of domination, authority, hierarchy is self-justifying: each must justify itself, and if it cannot, which is usually the case, then it should be dismantled in favor of greater freedom and justice. That seems to me the core idea of anarchism, deriving from its classical liberal roots and deeper perceptions—or beliefs, or hopes—about essential human nature. Libertarian socialism moves further to bring in ideas about sympathy, solidarity, mutual aid, also with Enlightenment roots and conceptions of human nature.

Both the anarchist and the Marxist visions have failed to gain ground in our own time, and in fact it could be argued that the prospects for the historical overcoming of capitalism appear to have been brighter in the past than they do today. If you do agree with this assessment, what factors can explain the frustrating setback for the realization of an alternative social order, that is, one beyond capitalism and exploitation?

Prevailing systems are particular forms of state capitalism. In the past generation, these have been distorted by neoliberal doctrines into an assault on human dignity and even the "animal needs" of ordinary human life. More ominously, unless reversed, implementation of these doctrines will destroy the possibility of decent human existence, and not in the distant future. But there is no reason to suppose that these dangerous tendencies are graven in stone. They are the product of particular circumstances and specific human decisions that have been well studied elsewhere and that I cannot review here. These can be reversed, and there is ample evidence of resistance to them, which can grow, and indeed must grow to a powerful force if there is to be hope for our species and the world that it largely rules.

While economic inequality, lack of growth and new jobs, and declining standards of living have become key features of contemporary advanced societies, the climate change challenge appears to pose a real threat to the planet on the whole. Are you optimistic that we can find the right formula to address economic problems while averting an environmental catastrophe?

There are two grim shadows that loom over everything that we consider: environmental catastrophe and nuclear war, the latter threat much underestimated, in my view. In the case of nuclear weapons, we at least know the answer: get rid of them, like smallpox, with adequate measures, which are technically feasible, to ensure that this curse does not arise again. In the case of environmental catastrophe, there still appears to be time to avert the worst consequences, but that will require measures well beyond those being undertaken now, and there are serious impediments to overcome, not least in the most powerful state in the world, the one power with a claim to be hegemonic.

In the extensive reporting of the recent Paris conference on the climate, the most important sentences were those pointing out that the binding treaty that negotiators hoped to achieve was off the agenda, because it would be "dead on arrival" when it reached the Republican-controlled US Congress. It is a shocking fact that every Republican presidential contender is either an outright climate denier or a skeptic who opposes government action. Congress celebrated the Paris conference by cutting back President Obama's limited efforts to avert disaster.

The Republican majority (with a minority of the popular vote) proudly announced funding cuts for the Environmental Protection Agency in order to rein in what House Appropriations Committee Chairman Hal Rogers called an "unnecessary, job-killing regulatory agenda"—or, in plain English, one of the few brakes on destruction. It should be borne in mind that in contemporary newspeak, the word "jobs" is a euphemism for the unpronounceable seven-letter word "pr---ts."

Are you overall optimistic about the future of humanity, given the kind of creatures we are?

We have two choices. We can be pessimistic, give up, and help ensure that the worst will happen. Or we can be optimistic, grasp the opportunities that surely exist, and maybe help make the world a better place. Not much of a choice.

Notes

Horror Beyond Description: The Latest Phase in the "War on Terror"

1. Katie Pisa and Time Hume, "Boko Haram Overtakes ISIS as World's Deadliest Terror Group, Report Says," *CNN*, November 19, 2015, www.cnn.com/2015/11/17/world/global-terror-report.
2. William Polk, "Falling into the ISIS Trap," Consortium News, November 17, 2015, https://consortiumnews.com/2015/11/17/falling-into-the-isis-trap.

The Empire of Chaos

1. Nick Turse, "Tomgram: Nick Turse, Success, Failure, and the 'Finest Warriors Who Ever Went into Combat,'" *TomDispatch*, October 25, 2015, www.tomdispatch.com/blog/176060.

Constructing Visions of Perpetual Peace

1. Noam Chomsky, *Who Rules the World* (Hamish Hamilton Ltd, 2016).

The Republican Base Is "Out of Control"

1. Andrew Cockburn, "Down the Tube," *Harper's*, April 2016, https://harpers.org/archive/2016/04/down-the-tube.

Trump in the White House

1. Dean Baker, *Rigged: How Globalization and the Rules of the Modern Economy Were Structured to Make the Rich Richer* (Center for Economic and Policy Research, 2016), deanbaker.net/books/rigged.htm.
2. Kristian Haug, "A Divided US: Sociologist Arlie Hochschild on the 2016 Presidential Election," *Truthout*, November 2, 2016, www.truth-out-org/opinion

/item/38217-a-divided-us-sociologist-arlie-hochschild-on-the-2016-presidential-election.

Global Warming and the Future of Humanity

1. Justin Gillis, "Flooding of Coast, Caused by Global Warming, Has Already Begun," *New York Times*, September 3, 2016, www.nytimes.com/2016/09/04/science/flooding-of-coast-caused-by-global-warming-has-already-begun.html.
2. Joby Warrick, "Why Are So Many Americans Skeptical About Climate Change? A Study Offers a Surprising Answer," *Washington Post*, November 23, 2015, www.washingtonpost.com/news/energy-environment/wp/2015/111/23/why-are-so-many-americans-skeptical-about-climate-change-a-study-offers-a-surprising-answer/?utm_term=.b9bd6860dfe2; Michael Roppolo, "Americans More Skeptical of Climate Change Than Others in Global Survey," *CBS News*, July 23, 2014, www.cbsnews.com/news/americans-more-skeptical-of-climate-change-than-others-in-global-survey.
3. Justin Gillis and Chris Buckley, "Period of Soaring Emissions May Be Ending, New Data Suggest," *New York Times*, December 7, 2015, https://mobile.nytimes.com/2015/12/08/science/carbon-emissions-decline-peak-climate-change.html.

The Legacy of the Obama Administration

1. On the latter matter, see Mary Ellen O'Connell, "Game of Drones," *American Journal of International Law* 109, no. 4 (2015): 889f.

The US Health System Is an International Scandal—and ACA Repeal Will Make It Worse

1. Daniel Schraad-Tischler, *Social Justice in the OECD—How Do the Member States Compare? Sustainable Governance Indicators 2011* (Gütersloh, Germany: Bertelsmann, 2011), news.sgi-network.org/uploads/tx_amsgistudies/SGI11_Social_Justice_OECD.pdf.
2. Walter L. Hixson, *American Settler Colonialism: A History* (Palgrave Macmillan, 2013), 2.

Index

Noam Chomsky

WHO RULES THE WORLD?

'As long as the general population is passive, apathetic, diverted to consumerism or hatred of the vulnerable, the powerful can do as they please and those who survive will be left to contemplate the outcome.'

In the post-9/11 era, America's policy-makers have increasingly prioritised the pursuit of power, both military and economic, above all else – human rights, democracy, even security. Drawing on examples ranging from expanding drone assassination programs to civil war in Syria to the continued violence in Iraq, Iran, Afghanistan, Israel and Palestine, Noam Chomsky examines the workings of imperial power across our increasingly chaotic planet.

'The world's greatest public intellectual' *Observer*

'One of the greatest, most radical public thinkers of our time. When the sun sets on the American empire, Chomsky's work will survive' Arundhati Roy

'[Chomsky is] the closest thing in the English-speaking world to an intellectual superstar' *Guardian*

NOAM CHOMSKY

ON ANARCHISM

On Anarchism is an essential introduction to Noam Chomsky's political theory.

This collection of Chomsky's essays and interviews provides a short, accessible introduction to his distinctively optimistic brand of anarchism. It sheds a much-needed light on the foundations of Chomsky's thought, specifically his constant questioning of the legitimacy of entrenched power.

Profoundly relevant to our times, *On Anarchism* is a touchstone for political activists and a book sure to challenge, provoke and inspire.

'Arguably the most important intellectual alive' *New York Times*

'Noam Chomsky is an inspiration all over the world – to millions, I suspect – for the simple reason that he is a truth-teller on an epic scale' John Pilger

NOAM CHOMSKY

POWER SYSTEMS

In this collection of conversations, conducted from 2010 to 2012, Noam Chomsky explores the most immediate and urgent concerns: the future of democracy in the Arab world, the implications of the Fukushima nuclear disaster, the 'class war' fought by U.S. business interests against working people and the poor, the breakdown of mainstream political institutions and the rise of the far right.

These interviews, conducted with David Barsamian, will inspire a new generation of readers and longtime Chomsky fans eager to understand the many crises we now confront, both at home and abroad.

'Noam Chomsky is a global phenomenon . . . he may be the most widely read American voice on foreign policy on the planet today' *New York Times Book Review*

NOAM CHOMSKY

MAKING THE FUTURE

'The fate of democracy is at stake in Madison, Wisconsin, no less than it is in Tahrir Square.'

Making the Future is a collection of essays from Noam Chomsky, one of our most vital and provocative voices of political dissent.

Taking up the thread from 2007's *Interventions*, these penetrating and compelling articles examine numerous topics, including the financial crisis, Obama's presidency, WikiLeaks and the on-going conflicts in the Middle East.

Restating and refining his commitment to democracy and finding inspiration in the popular uprisings of the Arab Spring, *Making the Future* is Chomsky's fiercely argued and timely comment on a fast-changing world.

'Chomsky is one of a small band of individuals fighting a whole industry. And that makes him not only brilliant, but heroic'
Arundhati Roy

NOAM CHOMSKY

OCCUPY

Since its appearance in Zuccotti Park, New York, in September 2011, the Occupy movement has spread to hundreds of towns and cities across the world. No longer occupying small tent camps, the movement now occupies the global conscience as its messages spread from street protests to op-ed pages to the highest seats of power. From the movement's onset, Noam Chomsky has supported its critique of corporate corruption and encouraged its efforts to increase civic participation, economic equality, democracy and freedom.

Through talks and conversations with movement supporters, *Occupy* presents Chomsky's latest thinking on the central issues, questions and demands that are driving ordinary people to protest. How did we get to this point? How are the wealthiest 1% influencing the lives of the other 99%? How can we separate money from politics? What would a genuinely democratic election look like? How can we redefine basic concepts like 'growth' to increase equality and quality of life for all?

'Noam Chomsky is an inspiration all over the world – to millions, I suspect – for the simple reason that he is a truth-teller on an epic scale'
John Pilger